Furthering Environmental Impact Assessment

Furthering Environmental Impact Assessment

Towards a Seamless Connection between EIA and EMS

Edited by

Anastássios Perdicoúlis
Engineering Department, UTAD, Portugal,
CITTA Research Centre, FEUP, Portugal

Bridget Durning
Oxford Institute for Sustainable Development,
Oxford Brookes University, UK

Lisa Palframan
University of Hertfordshire, UK

Edward Elgar
Cheltenham, UK • Northampton, MA, USA

Published by
Edward Elgar Publishing Limited
The Lypiatts
15 Lansdown Road
Cheltenham
Glos GL50 2JA
UK

Edward Elgar Publishing, Inc.
William Pratt House
9 Dewey Court
Northampton
Massachusetts 01060
USA

A catalogue record for this book
is available from the British Library

Library of Congress Control Number: 2012935292

ISBN 978 0 85793 327 0

Printed and bound by MPG Books Group, UK

Contents

Foreword

This book deals with an important topic. There is often a major discontinuity between the assessment of the potential impacts of a new project, via Environmental Impact Assessment (EIA), and the subsequent environmental management of its later construction, operational and decommissioning stages, via Environmental Management Systems (EMS). Yet in essence, both EIA and EMS can be seen as environmental protection tools with clearly complementary purposes. EIA seeks to anticipate and mitigate/enhance the impacts of proposed new projects at the planning and design stage; EMS helps organisations to effectively manage the day to day impacts through the full life cycle of such projects. The importance of the continuum seems obvious, and perceptive writers such as Holling (1978) spotted this many years ago – 'If assessment continues into the future, then prediction loses its status as a goal and assessment merges into environmental management.' Yet it rarely happens well in practice, and EIA runs the risk of being of only limited significance unless follow-up measures are carried out. Several writers and practitioners have tried to advance the continuum agenda, but there have been many barriers on the way.

This book deals with this important topic in an innovative way. It brings together an excellent mix of authors from academia, practice and a combination of both. They firstly take a theoretical perspective and explore and develop key concepts, principles and new ways of thinking about the impact continuum. They secondly provide a set of case study examples of innovative practice from several development sectors, and from different parts of the world. Early sections outline some of the barriers to date, but more importantly many of the advantages to be gained from overcoming such barriers. Such advantages include, for example, information efficiency gains, a focus on management capacity in EIA and not just prediction accuracy, and organisational learning and feedback from the project implementation stage into subsequent project planning and design. They also introduce ways forward such as Sustainability Management Systems (SMS), reflecting a more holistic and societal-based consideration of issues, risks and opportunities associated with project delivery. The case studies focus on key Infrastructure sectors including energy, transport, and water and waste management, with locations ranging from Scotland to Viet Nam, Iran, and especially Australia. Readers will find a wealth of innovative and practical tools contained in these case studies.

The book should be essential reading for the wide range of stakeholders involved in efforts to more sustainably manage a wide range of development activities. It covers well key highlights of current theoretical thinking and practical efforts by those active in the field, and hopefully it will be influential in meeting the editors' and authors' aim of driving forward the agenda to achieve an impact continuum.

Professor John Glasson,
Oxford Institute for Sustainable Development (OISD),
Oxford Brookes University

Acknowledgements

Our first word of thanks is to the authors of the chapters, who constitute the contemporary team of 'the impact continuum'. We would also like to thank the pioneers in the field, many of which have since followed other paths, for their own reasons. The fact that teams change may indicate some dynamics in the field, but it is good to know that the cause carries on.

We all wish to thank our respective families, for their 'support continuum' throughout the project. For their support we must also thank the Edward Elgar team – especially for their encouragement and patience. Coordinating a project like this, involving many people from different backgrounds and styles of work, is a bit of a challenge, and certainly a worthwhile experience.

Also, thanks to staff at Viridor for assistance with the Waste case study featured in Chapter 9.

Contributors

Dao Mai Anh finished her PhD in environmental compliance from the National University of Singapore in July 2011. Her main interest is in organisational behaviour and environmental management systems, with special focus on Vietnam-based corporations and their environmental performance.

Fiona Becker (CIWEM, PhD, MSc, BSc Hons, Registered EIA Assessor and Lead EMS Auditor with IEMA) Fiona is a Senior Environmental Specialist within ScottishPower Renewables (SPR) and is responsible for the development and operation of the company's Environmental Management System. The company's portfolio covers the development, construction and operation of onshore and offshore windfarms and it is a pioneer in the development of wave and tidal technologies. Fiona provides environmental management support to all these areas. She has over 10 years experience as an Environmental Professional and has worked in both the Consultancy and Utility Sector on Environmental Impact Assessment (EIA) for a range of developments (including wastewater, flood defence, landfill and renewable technologies) and environmental management. Fiona is currently the Chair of the Institute of Environmental Management and Assessment (IEMA) Scotland West Steering Group. Her Doctorate was awarded by Aston University, Birmingham, UK.

Clive Briffett (MSc, PhD) is an independent Environmental Consultant based in Oxford UK. He is a Visiting Fellow to Oxford Brookes University and a Visiting Tutor/Lecturer to the University of Oxford specialising in SEA/EIA, Ecotourism, Environmental Planning and Management. He has a PhD in Planning, Msc in Environmental Planning and Management, Msc in Architecture and Building Economics, and a Teaching Certificate in Further Education. **Note:** Clive passed away shortly before the conclusion of this book.

Martin Broderick is a Senior Technical Director at WSP Energy & Environment and a Registered Commissioner at the UK Infrastructure Planning Commission (IPC).[1] Martin has extensive (24 years) multi-sector experience in Environmental & Social Impact Assessment (ESIA). He has directed and managed large multi-disciplinary technical teams for Nationally Significant Infrastructure Projects in many sectors: Mining; Gas Pipelines; CCGT/CHP Power Plant; Thermal Nuclear Power Plant; Chemical Plant; Oil and Gas Exploration and Production onshore and offshore; Transport; Ports and Harbours; Telecoms; Commercial Developments; Underground Gas Storage. He has substantial international experience in Europe, Asia, FSU and Middle East. His clients have included a large number of FTSE 100 companies and he is familiar with regulators at the highest level in the UK and internationally. He is familiar with all relevant UK/EU legislation, World Bank, IFC Performance Standards and 'Equator Principles'. He has considerable experience in Environmental Due Diligence auditing for financial institutions and developers.

Bridget Durning (BSc, PhD, FGS, CGeol) is a senior research fellow with the Impact Assessment Unit (IAU) at the Oxford Institute for Sustainable Development (OISD), Oxford Brookes University, Oxford, UK. Bridget's background is in geology and she has over 25 years experience of working in the public and private sector in land development and planning as well as researching and teaching in higher education in the fields of environmental assessment and management and professional knowledge development. Bridget is a Chartered Geologist, Fellow of the Geological Society of London and a committee member of the Ireland/UK Branch of the International Association for Impact Assessment.

Neil Earnshaw (BSc Hons AIEMA) has worked in the environment and sustainability field since gaining an environmental degree in 1997. A member of the Institute of Environmental Management and Assessment since 2000, Neil's roles have included: business consultancy; training and auditing in management systems to a range of British and International standards for sustainability, environment, health and safety and quality. Neil moved from environmental consultancy to an environmental

[1] `http://infrastructure.independent.gov.uk/`

management role with Rolls-Royce. Neil established an independent sustainability consultancy in 2006 serving clients in the public and private sector including Rolls-Royce and the NHS. Neil joined the ODA in 2009 with responsibility for the sustainability management system implementation and for maximising the sustainability performance of the ODA's transport operations. Neil is currently working with both the ODA and LOCOG as the lead for sustainable transport for Spectators and Games Family transport respectively.

Karl Fuller (BA (Hons), MSc, MIEMA) has worked in the environmental assessment field for over 20 years. He helped to establish the organisation that was later to become the Institute of Environmental Management and Assessment. For some years Karl was responsible for technical services and advice provided by the Institute and contributed to the development, or was the principal author, of guidelines widely used in the UK and internationally. More recently Karl has worked in consultancy developing an approach to sustainability assessment for projects. He now works for the Environment Agency providing advice on the application of environmental assessment to flood risk management strategies, plans and projects.

Kevin House BA(Hons) qualified as a mechanical engineer before undertaking further academic studies at London Guildhall University. He obtained his first degree in Environmental Policy and Management and is currently working on a PhD in Environmental Management. Kevin lectured at Greenwich University's School of Environmental Sciences from 1996–2002 and continues to lecture on Environmental Management at the Royal School of Military Engineering (RSME). He joined the Environment Agency in 2003 as Environmental Strategy Manager for the Thames Estuary 2100 Project. Kevin joined the Environment Agency's National Environmental Assessment Service (NEAS) as Principal Environmental Project Manager in 2009.

Bruce Munro (BSc Msc MIEMA CEnv) is an environmental scientist with over ten years experience in the environmental sector, both in environmental consultancy and for the Environment Agency. Bruce has produced, managed and co-ordinated a wide range of environmental assessments, covering SEA, EIA, Habitats Regulations Assessment (HRA), ecology, hydrodynamics/geomorphology, water quantity and quality issues, diffuse pollution and catchment assessment, and the associated legislative and policy drivers. His role at the Environment Agency comprises the management of environmental risk and environmental assessment of some of the highest profile capital flood strategies and schemes in the country.

Lisa Palframan (MA (Oxon), MSc, MIEMA) is a senior lecturer in environmental management at the University of Hertfordshire, Hatfield, UK. Her research interests include EIA and the role of management systems in sustainability. She recently supervised a government-funded knowledge transfer project which developed a sustainability strategy for the Letchworth Garden City Heritage Foundation. Lisa also teaches environmental assessment and management to postgraduate level and is a member of the Institute of Environmental Management and Assessment's EIA Quality Review Panel. Prior to taking up her academic role, Lisa worked as an environmental specialist in industry and as an environmental assessment policy officer for a conservation charity.

Anastássios Perdicoúlis (BS, PhD) is an assistant professor at the engineering department of the University of Trás-os-Montes e Alto Douro (UTAD), Portugal, and affiliate researcher at the Research Centre for Territory, Transports and Environment (CITTA), Faculty of Engineering, University of Porto, Portugal. He has been working on planning methodology for twenty years, in fields such as spatial planning (including environmental planning), strategic planning, and public policy. More information is available at `http://www.tasso.utad.pt` or `http://www.fe.up.pt/~tasso`.

Jenny Pope Jenny Pope is Director of Western Australian consultancy firm Integral Sustainability, which provides consultancy and training services to Government and industry on the integration of sustainability concepts into project planning, strategy development and decision making processes at all levels. A chemical engineer by training, Jenny's practice builds upon her early career experience in industrial and corporate environmental management, particularly in the wastewater and the oil industries, in Australia and internationally. In 2007 she was awarded her PhD from Murdoch University for her research into the evolution of processes for the sustainability assessment of complex and strategic projects. In addition to her consulting practice, Jenny also maintains her interest and involvement in research for sustainability assessment by maintaining strong links with the academic sector. She works with Masters level students at two universities (Cambridge University in the UK and North West University, South Africa) as a tutor and dissertation supervisor. She also regularly gives guest lectures and provides guidance to research students upon request; participates actively in the International Association for Impact Assessment and attends other conferences; maintains currency with the literature and reviews papers for journals; and prepares conference presentations and papers through a process of continual reflection upon her professional practice.

Behzad Raissiyan has been working as Environmental Advisor for a major oil development project for the past two years, with specific responsibility for implementation and follow-up of the EIA commitments and establishment of an EMS. Prior to his current role, Behzad was a freelance environmental advisor, working on EIA for industrial and tourism developments and strategic environmental assessment for regional development plans. Behzad is a registered third party lead EMS auditor and senior tutor for ISO 14001, ISO 9001 and OHSAS 18001. He holds a BSc in Metallurgy Engineering and an MSc in Environmental Planning and Management and is a member of the International Association for Impact Assessment.

Luis E. Sánchez teaches environmental planning and management at Escola Politécnica, University of São Paulo (USP). He graduated in mining engineering and in geography, both at USP, and obtained his PhD in economics of natural resources from the Paris School of Mines, France. He was invited lecturer at the University of Montreal, Canada and research fellow at Murdoch University, Australia. He authored several articles and book chapters on environmental assessment and management and a textbook on environmental impact assessment, published in Portuguese and translated into Spanish. He has been active in the International Association for Impact Assessment since 1992.

Jane Scanlon is a specialist in sustainability management systems and governance frameworks for infrastructure project delivery. Her clients have included the New South Wales Transport Construction Authority, Gold Coast Rapid Transit, Gold Coast City Council and Energised Alliance (Ausgrid). She is a trained CEEQUAL assessor and an active representative of and adviser to the Australian Green Infrastructure Council. Jane recently submitted a PhD thesis on rail project sustainability management at the University of Western Sydney, where she examined the sustainability management approaches of the UK Olympic Delivery Authority transport division and the Sydney Metro Authority. During her research programme, she established a global network of infrastructure sustainability specialists, and has published numerous industry and academic articles on the topic.

Claire Vetori (BSc(Hons), MIEEM, CEnv) is an ecologist by training having studied Pure and Applied Ecology at the University of Sheffield, graduating in 1994. After volunteering part-time as an assistant ranger at a local country park, Claire entered into environmental consultancy. During her 15 years working for several private consultancies in London and the South East, Claire specialised in Environmental Impact Assessment work. More recently, since joining the Environment Agency in early 2010, Claire has worked on Strategic Environmental Assessments and become a specialist in the requirements of the Water Framework Directive and how it applies to the flood and coastal risk management activities undertaken by the Environment Agency.

Hudson Worsley (BA, BBus, MEM) is a sustainability professional with a mix of consulting, project and corporate experience. He has a background in mainstream management consulting and spent over five years working for Manidis Roberts with clients on complex development and infrastructure projects in residential property, energy, water and transport. He was the environmental sustainability manager for the Sydney Metro Authority and was responsible for establishing the documentation framework that made up the organisation's sustainability management system. Hudson is now working in the property development sector as the national sustainability manager, Residential Communities for Stockland, where he leads the sustainability strategy development and implementation across Stockland's portfolio of residential communities (Australia's largest).

1. Introduction

Bridget Durning, Lisa Palframan, and Anastássios Perdicoúlis

Environmental impacts of investment projects – for instance, petroleum drilling, mining, or highways – are currently studied and mitigated from two distinct perspectives: before and after the project implementation, with environmental impact assessment (EIA) and environmental management systems (EMS) being the main instruments on the respective sides. This double-perspective creates a discontinuity in the way environmental impacts are dealt with but there is no significant reason why this situation should remain sub-optimal as EIA and EMS are both environmental protection tools with complementary purposes: whilst the goal of EIA is to anticipate and mitigate the environmental impacts of proposed new projects at the planning and design stages, an EMS can help organisations to effectively manage the day-to-day environmental impacts arising during the construction, operation and decommissioning of such projects. By supporting a systematic approach to the identification and evaluation of impacts, both tools can ensure that resources are focused on those impacts deemed to be 'significant'. Used effectively and in an integrated manner, they can ensure the main environmental issues are identified at an early stage in project planning and are systematically addressed throughout the project life cycle.

The issues with this double-perspective, but also the potential benefit of closer connection, have been perceived by both academics and practitioners at various points over the last forty years. There have been both theoretical and practical attempts to link EIA with environmental management (EM) almost from the point when EIA was first formally enacted. In this book we aim to drive forward the agenda of achieving an 'impact continuum' by gathering and presenting highlights of current theoretical thinking and practical efforts as expressed by those active in the field today in linking EIA and EMS.

We are aware that EIA is only one of a number of tools for assessing impact of both projects, plans and policies including Sustainability Appraisal (SA), Strategic Environmental Assessment (SEA), Health Impact Assessment (HIA), Territorial Impact Assessment (TIA), Regulatory Impact Assessment (RIA), Equality Impact Assessment (EqIA). In particular we aware that Social Impact Assessment (SIA) is often undertaken separately to EIA. In this book we are particularly focusing at project level continuum.

We start by setting the context for book: section 1.1 of this chapter provides a brief summary of the key elements of EIA and EMS and the perceived problems with their integration. Section 1.2 reflects on how the 'impact continuum' has been presented in academic literature over the last forty years. Finally, section 1.3 provides a brief summary of each of the chapters in the book, and an outline of the book's readership.

1.1 Key Elements of EIA and EMS

1.1.1 EIA

EIA is a systematic process which has been practised for more than 40 years. It is now carried out in more than 100 countries around the world, as well as by many bilateral and multi-lateral funding agencies (Petts, 1999). Its purpose is to predict, assess and evaluate the environmental impacts of major development projects before a decision is made on whether they should be consented. Such projects may include airports, highways, pipelines, mines, power stations and large scale urban development projects. The main principles of EIA can be described as an aid to decision making (ensuring that the decision is better informed), an aid to the formulation of development actions (by anticipating environmental challenges at an early stage in the project design) and an instrument for sustainable development (through the avoidance of environmental damage) (Glasson et al., 2005).

Although EIA is practised internationally, there are a number of steps in the procedure which are commonly followed up to the consenting decision, including:

- screening, required to decide – rather technically – whether or not EIA is needed

- scoping, required to establish the main issues requiring consideration, any alternatives that should be considered and the methods/techniques that should be employed to investigate the identified impacts
- prediction and assessment of impacts, drawing on baseline data to understand the likely changes in the environment that may arise from the project
- evaluation of impacts in terms of their likely significance
- development of mitigation and monitoring measures, to eliminate or reduce any negative impacts identified and establish whether the predicted impacts are accurate

One significant output of the EIA process is a document which communicates the process and findings of the EIA, variously called an Environmental Impact Statement (EIS), Environmental Statement or EIA Report. This document, together with the application for project consent, is submitted by the developer to the consenting authority to inform the decision about whether and if so, how, the project should proceed. An important element of EIA is public participation; individuals, communities and organisations likely to be affected by the development should have the opportunity to access the environmental information (for example through the widespread dissemination of a non-technical summary of the EIS) and have their views on the development taken into account in the decision making process. While many countries require public consultation only at the end of the EIA process, it is considered good practice to actively involve the public much earlier and at regular intervals during EIA, bringing benefits including increased opportunities for modification of the development and improved relationships between the parties (IAIA, 2006).

The formal and mandated EIA procedure in some jurisdictions ends with the consideration of the EIS and the granting of consent. In many countries however it is common for there to be formal, regulated 'follow-up' required by EIA or other environmental protection laws which may encompass permits, standards, surveillance and enforcement of compliance (Morrison-Saunders and Arts, 2004a). These approaches enable the regulator to check that the development is proceeding in accordance with the EIA and/or planning documentation, predicted impacts are controlled and unanticipated impacts are avoided. Within countries where formal regulation exists and where the developer is left to self-regulate, the implementation of an EMS as part of the follow-up strategy allows the risks identified during EIA to be managed during the construction and operation of the project.

1.1.2 EMS

An EMS is a voluntary measure which enables an organisation to systematically identify and manage its environmental impacts, to improve environmental performance. It is based on the 'plan-do-check-act' cyclical management model also described as Total Quality Management (Netherwood, 1994). It enables the environmental policy that many organisations adopt to be actively implemented through the ongoing identification of risks and impacts together with the achievement of actions to enable continual improvement. Organisations that choose to adopt this approach are motivated by the possibility of improved risk management, regulatory compliance and enhanced reputation. Some organisations operate a self-certified EMS, others choose to have their EMS externally assessed and certified against a recognised standard such as the international standard ISO 14001. An EMS usually includes the following components (based on ISO (2004)):

- an environmental policy statement, which sets out the organisation's commitment to managing its environmental issues
- a register of significant environmental aspects (environmental interactions) and impacts (changes in the environment resulting from these interactions). These are often established through carrying out an environmental review, a strategic process identifying the main environmental impacts of the organisation's operations, products and services.
- a register of the main laws and regulatory requirements applicable to the organisation
- an environmental improvement plan which includes objectives and targets to address the significant aspects
- documented, implemented procedures which ensure satisfactory provision is made in areas such as resourcing, communication, staff training and competence, emergencies and document control
- the monitoring and auditing of significant environmental aspects, compliance with the EMS and achievement of the objectives and targets. Actions

are needed to address the issues found in an effective manner

- a regular, documented review of the EMS in which senior management establish the progress made and assess further action needed (e.g. changes to environmental aspects, addition of laws to the legal register, new targets) in light of any internal or external changes

In some cases an organisation will chose to report publicly on their environmental performance. This is a requirement of some standards such as EU EMAS (European Communities, 2011) but is only an option under other schemes. EMS standards specify what must be achieved in terms of management rather than performance; for example they do not specify limits on resource use. This means they can usually be followed by public or private sector organisations operating in any industry in a way that gives the organisation considerable discretion over the issues to be addressed and the rate at which progress is made. In theory, by putting in place mechanisms to integrate environmental concerns into day-to-day operations, an organisation can enable environmental performance improvements to be realised, but some researchers have argued that there is limited agreement on what is meant by 'performance' in an EMS context and why performance improvements should result (Nawrocka and Parker, 2009). However, it is clear that where organisations need to take a systematic approach to address the environmental impacts of their ongoing operations, EMS models provide an appropriate framework.

1.1.3 Open Challenges of the EIA–EMS Interface

A comparison of EIA and EMS (see Table 1.1) demonstrates many differences between the origins, process and application of the two tools, although it is evident that their goals of environmental protection are similar. This suggests that both tools are approaching a similar problem from different perspectives and although there may be practical issues to overcome in applying them to the same project, closer linking of the tools where they are to be applied to the same development project would be beneficial. Some of the practical issues identified in previous studies are identified in Table 1.2.

Table 1.1 A comparison of EIA and EMS elements and requirements; structure based on Eccleston and Smythe (2002)

Characteristics	Typical EIA Process in the UK	ISO14000-consistent EMS
History	EIA Directive implemented in 1988, although many EIAs prepared voluntarily or at the request of local authorities before this time (Glasson et al., 2005)	First British Standard in 1994; ISO standard 14001 adopted in 1996
Goal	To provide environmental protection by ensuring environmental considerations are integrated into decision making for major developments	To provide environmental protection by identifying impacts; through systems for continuous improvements, to reduce these impacts
Scope	Proposed projects listed in Annex 1 or 2 of the EIA Directive deemed likely to have significant environmental effects	Defined in the ISO 14001 certificate as applying to all activities within the specified site or organisation
Organisations directly involved	Organisations intending to apply for permission to develop a project specified on the annexes of the EIA Directive must undertake EIA; majority private sector, occasionally public sector	Voluntary approach open to any organisation that wishes to manage its environmental impacts in accordance with a recognised standard
Personnel directly involved	Usually external consultants prepare the Environmental Statement; some are prepared in-house	Consultancy support may occasionally be sought in setting up the EMS but they are usually implemented by in-house personnel with day-to-day responsibility for environmental issues
Public participation	Public consultation on the Environmental Statement is mandatory; public may also be involved at earlier stages	Organisation must 'decide whether to communicate externally about its significant environmental aspects and shall document its decision' (ISO, 2004)

(Continued on next page)

Characteristics	Typical EIA Process in the UK	ISO14000-consistent EMS
Planning	An optional scoping stage allows for the EIA to be planned in terms of the issues to be assessed and the methods by which this will be achieved; a Scoping Report may be issued to aid consultation at this stage	'Scoping' achieved through carrying out an Initial Environmental Review which identifies and evaluates significant aspects and impacts; an Environmental Policy is produced based on the issues identified and the commitments the organisation wishes to make
Documentation	Environmental Statement summarising the findings of the EIA must accompany the planning application (information requirements set out in Schedule 4 of UK EIA regulations); a Non-Technical Summary is usually published separately to aid public consultation; neither are actively maintained, being produced for a set point in time (the consenting process)	Environmental Policy developed, maintained and communicated; evolves according to changing circumstances; various procedures required to be documented and maintained, e.g. on legal requirements and communication; documents are required to be controlled
Significance	For the purpose of screening, significance criteria are set out in Schedule 3 of the UK EIA regulations and are broadly classed as characteristics of the impact, characteristics of the development and location of the development	For the purpose of evaluating the significance of aspects associated with the organisation's activities, suggested significance criteria are set out in ISO 14004 and broadly include environmental impact, legal requirements and stakeholder concern
Mitigation	UK EIA regulations require the Environmental Statement to provide 'A description of the measures envisaged to prevent, reduce and where possible offset any significant adverse effects on the environment'	EMS provides a system for ensuring that mitigation measures are implemented, for example the activity they relate to may be identified as a 'significant aspect'; they may be formally monitored, audited and reviewed

(Continued on next page)

Characteristics	Typical EIA Process in the UK	ISO14000-consistent EMS
Monitoring	No legal requirement to implement monitoring, nor to undertake related 'follow-up' activities, although individual proponents may be obliged to do so through the consenting authority making planning conditions	Requirement to maintain and implement a procedure to monitor and measure the significant impacts of the operation; in practice this includes monitoring of resource used and regular internal audits; a management review is required 'at planned intervals'

One particular issue is the motivation of organisations for using each of these tools. With EIA being a legal requirement for certain projects in many countries, organisations have no choice but to carry it out in order to obtain planning consent. In contrast, EMS is not usually legally required, although in some cases sectoral requirements make the implementation of an EMS compulsory for companies carrying out certain activities. For example, a recommendation by OSPAR (OSPAR Convention for the Protection of the Marine Environment of the North East Atlantic) requires Contracting Parties to ensure that all offshore oil and gas operators have an internationally recognised EMS (OSPAR Commission, 2003). Arguably, the voluntary nature of EMS limits its uptake amongst companies that have put their proposed projects through the EIA process, even where there may be efficiency and environmental protection benefits to using an EMS during construction and operation of the project. Furthermore, in many jurisdictions the system for the planning of new developments (in which EIA has a primary role) and the system for regulating pollution from industry (in which EMS has a primary role) are different. This often results in at least two separate consenting applications, sometimes to two different authorities, one for planning permission and one for an environmental permit. In the European Union for example, there are separate Directives on EIA (85/337 as amended) and on Integrated Pollution Prevention and Control (2008/1/EC). The separation of the consenting systems may widen the gap between the tools that support favourable environmental protection outcomes.

Relevant to many of the concerns highlighted in Table 1.2 are several concepts which present difficulties in translation from EIA to EMS and vice versa. The first of these is 'environmental aspect', defined by ISO (2004) as an 'element of an ation's activities or products or services that can interact with the environment'. Sánchez and Hacking (2002, p. 28) describe an aspect as 'the linkage between an activity, product or service and their environmental consequences, or impacts'. An organisation wishing to implement an EMS that meets the requirements of ISO 14001 must identify their aspects and evaluate their significance. The term 'aspect' is not routinely used in EIA, instead it is the case that impacts are usually classified according to the biophysical features they affect. Morris and Therivel (2009, p. 534) describe these features as 'environmental components', a term defined as 'the aspects of the natural or man-made environment (e.g. people, landscape, heritage, air, soils, water, ecosystems) that may be significantly affected by a proposed project, and are individually assessed in an EIA. They are receptors, but can also include many of these, e.g. individual species or buildings'. Thus on the rare occasions that the term 'aspect' is used in an EIA context, it may be interpreted as an environmental feature which a business, through its proposed development, may affect. In contrast, the term 'aspect' in EMS refers to the connection between the business and the environment, rather than the environmental receptor specifically (ISO, 2004).

Table 1.2 Perceived barriers to linking EIA and EMS according to existing work; based on Palframan (2010)

Type of Barrier	Example	Reference
Legal and policy framework	Different consenting regimes for planning and environmental protection (implied)	(Environmental Resources Management, 2004)
	Potential overlap in requirements leading to inefficiencies	(Eccleston and Smythe, 2002)
	Voluntary basis of EMS providing little incentive for uptake	(Slinn et al., 2007)
Process/technical issues	Complexities of site ownership and occupation	(Slinn et al., 2007)
	Time lag between EIA being carried out and detailed design of the project	(Ridgway, 2005)
	EMS orientated towards day-to-day activities, environmental implications of new development not considered	(Marshall, 2004)
	Limited number of practitioners specialising in both tools	(Sánchez and Hacking, 2002; Marshall, 2004)

(Continued on next page)

Type of Barrier	Example	Reference
Practitioner issues	Different personnel undertaking EIA and EMS for any given project	(Ridgway, 2005; Sánchez and Hacking, 2002)
	Public debate around new developments centred on whether or not to grant consent, not on mitigation	(Sánchez and Hacking, 2002)
	Companies consider EMS to be outside the normal scope of operational activities	(Marshall, 2004)
Proponent and stakeholder attitudes	EIA viewed by proponents as a bureaucratic step rather than a useful process to aid the delivery of the project	(Sánchez and Hacking, 2002)
	Reluctance of proponent to put resources into operational management before the outcome of the application is known	(Slinn et al., 2007)

The concept of 'significance' is another which presents difficulties in the relationship between EIA and EMS because it is used differently with each of the two tools, rather than being unfamiliar to one. ISO 14001 (ISO, 2004) explains that a significant aspect (of the activity, services or products of the organisation) is one which has significant impacts on the environment (although the organisation itself makes the judgement on the threshold above which the impact is deemed significant) and that the EMS should be structured around addressing those significant aspects. The guidance (ISO, 2004) makes clear that the organisation itself must establish a documented method and criteria for evaluating significant aspects, perhaps taking into account environmental matters, legal issues and stakeholder concerns. An organisation operating an ISO 14001 EMS usually creates and maintains a register of significant environmental aspects and associated legal requirements. The identified aspects should be used in determining the programme of objectives and targets with resources prioritised towards the most significant aspects. This programme is central to the achievement of continual improvement in environmental performance especially in companies which set ambitious goals (Brouwer and van Koppen, 2008).

The concept of significance with regards to EIA has been described as 'one of the most complex, contentious, and least-understood aspects of EIA systems across the globe', particularly due to its dynamism through the project cycle (Wood, 2008, p. 23). Significance is often used during the screening process as a way of determining which projects should be subject to EIA (those that are determined, on the basis of limited information at this early stage, as likely to have significant effects). Assessment should result in identification and confirmation of impact significance, with mitigation measures being put forward to reduce significance where possible. As with EMS, the determination of significance helps to ensure that limited resources are prioritised so that the greatest environmental impacts are identified more fully during assessment and that they are then avoided or reduced during implementation of the project. There is no universal agreement on how a 'significant' impact is defined in EIA, although it is common to see terms such as 'slight', 'moderate' etc. used to describe impact significance within EIA reports (Wood et al., 2007). As with EMS, significance can be seen as a value judgement, open to interpretation. However, it should be underpinned by rigorous evidence which substantiates the judgement. For some issues such as air quality and noise, evaluation of significance can be carried out against recognised and agreed environmental quality standards which should make judgements more consistent. Arguably EIA is better positioned than EMS to present a more robust judgement on significance where such standards do not exist because of the requirement to involve a wider range of stakeholders – although ISO (2004) advises that a company take account of stakeholder views in determining significance, it is acceptable for companies to show very limited evidence of external stakeholder involvement. However, since most EIA systems only require consultation once the EIA report is completed, the significance judgement may still be developed without reference to stakeholder views.

1.2 Impact Continuum Over the Last Forty Years

The two processes, which we are proposing should be a continuum rather than separate entities had their origins in initiatives to address pollution and environmental degradation, originally brought to the fore through the environmental movement of the 1960s.

EIA in a formal sense first appeared in the USA in 1969 through the National Environmental Protection Act (NEPA) (Wood, 2003). Legislation requiring the implementation of EIA gradually spread across the developed world over the next 20 years. E(S)IA (recognising that in some legislation social impact assessment is integrated and others it is separate) is now a mature process is most developed countries. EMS had a slightly later birth, with formalised systems not appearing until much later in 1992 through the publication of the British Standard (BS7750) and subsequently of the ISO 14001 in 1996. This was followed by the European Union's Eco Management and Audit (EU-EMAS) scheme in 1995. Since then various sectoral and national level schemes have also been introduced, for example the Acorn Scheme in the UK for small and medium sized enterprises (IEMA, 2010) and the National Framework for Environmental Management for the Australian agricultural sector (Australian Government Department of Agriculture Fisheries and Forestry, 2009).

Holling (1978) is notable for being a very early piece of work that proposed the environmental assessment and management should be a continuum:

> If assessment continues into the future, then prediction loses its status as a goal and assessment merges into environmental management. Prediction and traditional 'environmental impact assessment' suppose that there is a 'before and after' whereas environmental management in an ongoing process (Holling, 1978, p. 133).

More innovative work continued through the 1980s with exploration of the need for follow-up and monitoring in EIA to extend it into EM – see Bailey (1997) and references therein.

Subsequent to the arrival of formalised systems for environmental management in the 1990s, work looking specifically at EIA–EMS began to appear in the academic literature, notably Eccleston (1998) and an innovative special edition of the Journal of Environmental Assessment, Policy and Management (JEAPM) in 1999. Eccleston's work offered a conceptual framework which showed the potential for synergy between EMS and the EIA process (within NEPA) and proposed that integrating the two could 'lead to more effective planning and enhanced environmental protection, while streamlining compliance' (Eccleston, 1998, p. 10). The theoretical benefits from a combination of EIA and EMS are also presented by Ridgway (1999).

Some practical examples also start to appear in the literature with Barnes and Lemon (1999) providing an example of the use of an Environmental Management Plan within a Canadian road development. However, others also began to identify that in some industries EIA is not enough to identify and mitigate significant impact and more is needed, like EMS – for instance, McKillop and Brown (1999) who considered an example from the extractive industries. This period also saw other actors entering the arena and Rees (1999) charts the rise of IFC World Bank Equator Principles.

Sheate (1999) in his editorial to JAEPM special edition, observed that the abundance of academic articles on the integration of EIA with other tools which appeared at this time showed an expanding area of academic interest and this appears to have continued over the last decade with a number of papers that have appeared in the academic literature at various points – for instance, Eccleston and Smythe (2002), Sánchez and Hacking (2002), Vanclay (2004), Marshall (2004), Ridgway (2005), Broderick and Durning (2006), Slinn et al. (2007), Perdicoúlis and Durning (2007), Cherp (2008), Varnäs et al. (2009), and Lundberg (2011).

The published academic work contains both considerations on theoretical integration and a few examples of actual integration, such as Marshall (2004) and Broderick and Durning (2006). There was also (and continues to be) focus in the literature on the need for and role of 'follow-up' to EIA. According to Arts et al. (2001), environmental management is one of the four key activities of EIA follow-up (which also include monitoring, evaluation and communication). Morrison-Saunders and Arts (2004a, p. 4) describe management as: 'making decisions and taking appropriate action in response to issues arising from monitoring and evaluation activities' and as such can be considered a key to co-ordinating other follow-up activities.

Examples of the use of environmental management plans have appeared at various points in the literature – for instance, Barnes and Lemon (1999), Broderick and Durning (2006) – and in 2008 IEMA in the UK published good practice guidance on their use (IEMA, 2008). There is some variation in the extent of use of EMPs depending on their sector and the driving force, which is explored further in Chapter 4 of this book.

As mentioned previously, ongoing development has expanded the range of impact assessment tools beyond those purely focused on the project delivery aspect. Strategic environmental assessment (SEA) has developed as a tool which complements EIA. It enables environmental considerations to be integrated into the preparation of strategic actions such as policies, plans and programmes.

These actions often set the planning framework for the types of capital projects that are likely to require EIA at a later stage. The advantages of this tool include its early application in the planning cycle (meaning that it can influence the types of projects, not just the details of the projects), its ability to deal with cumulative impacts and the possibility for more meaningful consideration of strategic alternatives (Thérivel, 2004).

On the EMS side, researchers have developed approaches to expand the scope of EMS towards other issues of sustainability (Azapagic, 2003; Esquer-Peralta et al., 2009). A Sustainability Management System (SMS) may be viewed as a mature standard EMS that takes a broader sustainability perspective (Emilsson and Hjelm, 2009). Depending on the characteristics of the organisation, such issues may include workplace conditions, the influence on the local community and support for the local economy. There is currently no ISO 14001 equivalent for SMS, although draft standard ISO 20021 setting out a specification for SMS in events management is in preparation and due to be published in 2012 (ISO, 2010b). Nonetheless, practitioners are beginning to report on their experiences of SMS – for instance, MacDonald (2005), Suff (2011). Chapter 11 explains SMS and provides guidance on developing SMS in the context of projects that may be subject to EIA, whilst Chapter 12 examines sustainability issues within the context of two case study projects.

1.3 About the Book

1.3.1 Content

As is the nature of a subject which is firmly rooted in practice, many of those 'pioneers' in the early consideration of EIA–EMS, have now moved on into other areas of EIA practice and new practitioners have moved in. It is probably a reasonable assumption that there is a lot of experience of the impact continuum in practice as many companies have EMS (more than 223,000 companies have ISO 14001 according to latest figures from ISO (2010b)) and because EIA also affects those same companies when undertaking major development. However, it is difficult to know whether and if so, how, companies actually integrate EIA beyond the limited literature to date. The time is now right for a book which sets the current picture and explores how EIA–EMS integration has evolved and continues to evolve. This book therefore contains a number of case studies from a range of sectors to provide examples of current practice. A summary of each chapter follows:

Chapter 2 explores the relationship between project environmental planning and management and the contribution of information and knowledge management (I&KM) to the achievement of environmental protection outcomes. It outlines the basis for I&KM in the EIA process and describes the typical contents of the contemporary environmental manager's toolbox. It goes on to review the integrative use of selected tools to achieve better integration between project planning and management and discusses how environmental information generated during project implementation and management can be transformed into knowledge to improve the assessment of future projects.

Chapter 3 presents an analysis of the EIA and EMS processes, featuring documents and tasks as the main elements. Considered in a document-only, task-only, or an integrated view, the analysis provides two alternative perspectives of the processes. The 'action perspective' of the integrated view provides the most complete and conservative framework for the EIA–EMS link. On the contrary, the 'milestone perspective' of the integrated view provides a more tangible core sequence, and hence an opportunity for a more flexible (re-)definition of tasks – a deviation from current standards which may be revolutionary for either process, with the advantages and disadvantages of any revolution.

Chapter 4 There is no formal requirement for an environmental management plan (EMP) within EIA legislation in the UK, but they are increasingly being used as a way of ensuring that mitigation measures proposed in the environmental statement (ES) are implemented on the ground. This chapter considers the global origins of the concept of EMP. It describes their current usage and briefly considers their effectiveness in delivering environmental protection through reference to a case study. It concludes by reflecting on how EMPs can form the key link between EIA and EMS and how this link can be further enhanced.

Chapter 5 Appropriately employed, environmental and social impact assessment (ESIA) is a key integrative element in environmental protection, but is only one element of that policy toolbox. Other elements include the monitoring and evaluation of the impacts of a project (which has been subject to ESIA) and the subsequent management of the environmental performance of that project. It has been suggested by some authors that ESIA has little value unless follow-up is carried out: without it the process remains incomplete and the consequences of ESIA planning and decision making are unknown. Through an exploration of the many drivers for ESIA development and differing methodological approaches, this chapter proposes that the most effective ESIA systems are those that use follow-up processes to tangibly link ESIA and EMS.

Chapter 6 Using the industrial sector within Vietnam as a case study, this chapter explores the potential of utilising Environmental Management System (EMS) in association with Environmental Impact Assessment (EIA) to develop a framework to guide the environmental law and regulation compliance behaviour of the sector. It first presents the background to the empirical study on which the chapter is based. It presents a country overview followed by outlining current EIA and EMS practice in Vietnam, the determining factors in compliance with environmental laws and regulations and concludes with a discussion of the role of EMS.

Chapter 7 Raissiyan and Pope discuss two case studies from the oil sector in Iran, in which practical steps were taken to enhance the linkages between EIA and EMS, drawing on one of the author's experiences both as an EIA consultant and EMS practitioner. They discuss the barriers and challenges which exist in linking these tools, particularly in developing countries, and propose solutions which would enable project teams to work more effectively towards integrated solutions.

Chapter 8 An industry EMS practitioner discusses how EIA and EMS have been applied together through the life cycle of a wind farm located in Scotland. Practical examples are given showing how links have been made through staff working together to apply the two tools during the development, construction and operational stages. The role of the regulator in providing an additional knowledge feedback loop is emphasised.

Chapter 9 focuses on the planning and pollution control of waste management infrastructure and discusses the use of EIA and EMS with reference to an extension to a landfill site located in the UK. The company which runs the site have developed both a corporate and site-based approach to ensuring that EIA and the Business Management System (incorporating an EMS) are linked and that information is effectively shared to meet environmental requirements. The chapter discusses how this approach could be applied by companies in other sectors.

Chapter 10 discusses how the Environment Agency in England and Wales uses an Environmental Action Plan to link EIA and the EMS during flood risk management projects. The chapter describes the involvement of contractors in risk management and discusses how monitoring from existing projects informs good practice elsewhere.

Chapter 11 Scanlon and Pope introduce the concept of a Sustainability Management System (SMS), which can allow for social and environmental impacts and opportunities to be addressed more holistically than a traditional EMS. They propose that SMS be used to integrate consideration of environmental and societal issues at all stages of project planning and delivery – including in design, procurement and construction. They also provide guidance on how to develop and implement a SMS.

Chapter 12 builds on the discussion of SMS presented in Chapter 11 by exploring case studies in the transport sector; one the Sydney Metro Authority (SMA) in Australia and the other the UK Olympic Delivery Authority Transport Division (ODAT) in the UK. The chapter discusses how SMS has been used to embed sustainability during the planning and delivery of major infrastructure projects and how such an approach can overcome some of the inhibitors to project sustainability.

1.3.2 Readership

The book should appeal primarily to academics related to departments and/or courses such as environmental planning, environmental engineering, or environmental management, and to three groups in particular: (a) researchers who study and develop methods and applications of EIA–EMS integration; (b) lecturers who teach in any of the two 'components' (EIA or EMS) or the full 'continuum'; and (c) undergraduate and post-graduate students – the former as a first exposure to the concepts in an integrating perspective, and the latter facing practical or theoretical problems to solve.

The book should also appeal to 'avant-garde' project proponents and practitioners (for instance, consultants), who often work together. The book aims to provide a stimulus and practical examples for them to also experiment with the various aspects of EIA–EMS integration, so that soon we can evidence a wave of efforts in the same direction – perhaps reinforcing the trend of the 'impact continuum'.

2. Information and Knowledge Management

Luis E. Sánchez

Timely, accurate and purposeful information is essential to environmental impact assessment (EIA), as it provides necessary inputs to project planners and decision makers. The same is true for managers in charge of constructing, operating or decommissioning a project whose approval was supported by EIA. But what is the connection between information generated in the planning and subsequent phases of a project's life span?

Managers are supposed to comply with terms and conditions of approvals and licenses which in turn derive from impact assessment. Thus, mitigation and compensation of negative impacts, enhancement of positive ones, risk reduction measures, emergency action plans and monitoring are components of management that originate in the assessment phase.

In practice, however, assessment teams are different from the construction crew which in turn hands over the facility's keys to a new operations team. Can we ensure the integrity of relevant information as it flows through so many heads? In fact, it is not information that matters, but knowledge. Know-how, know-what, know-when and especially know-why are perhaps the key aspects to be transmitted as a project is conceived, assessed, modified, reassessed, built and operated. It is different for a construction manager to implement a particular mitigation measure because someone told him he has to (even because of contractual conditions) or to implement the very same measure because he knows why it is part of the project approval conditions – the result of a negotiation with the community, for instance.

Conversely, as information goes downstream from project planning to management, it can also go back from managers to planners to feed the assessment of new projects. Data collected in monitoring programmes, supervision or audits can be processed to generate knowledge useful to the assessment of forthcoming projects.

This chapter will explore the relationship between project environmental planning and management and the contribution of information and knowledge management (I&KM) to the achievement of environmental protection outcomes. Section 2.1 will provide a basis for I&KM in the EIA process. Section 2.2 outlines the contents of the contemporary environmental manager's toolbox, selecting those more relevant to project management. Section 2.3 reviews the integrative use of selected tools to achieve better integration between project planning and management. Section 2.4 deals with managing the project for achieving environmental outcomes and Section 2.5 discusses how environmental information generated during project implementation and management can be transformed into knowledge to improve the assessment of future projects.

2.1 Information and Knowledge in the EIA Process

EIA is a 'macro process' in support of decision making, but project implementation involves a series of 'micro decisions'. Decisions derived from the process include project siting, technology, mitigation and enhancement, among others. Such decisions are essentially dealt with by the EIA system adopted by each jurisdiction, usually associated to some form of permitting or licensing establishing legally binding – and presumably enforceable – terms and conditions of approval. Micro decisions, on the other hand, relate to both detailed design – prior to construction but often after project licensing – and on-the-site problem-solving. As a consequence of those micro decisions, actual impacts can be smaller than predicted, e.g. less vegetation clear cut achieved, but can also be larger than expected.

Those micro decisions must be informed by and consistent with EIA outcomes, i.e. mitigation and enhancement described in the environmental impact study and EIA terms and conditions listed in the project approval document. This section discusses how to ensure that information and knowledge originated in EIA do not get lost, forgotten or misunderstood when it is time to make such micro decisions. In other words, can information and knowledge generated in the assessment phase be properly transferred to the management phase? How should knowledge be stored and retrieved for management purposes?

A practical answer to those questions, adopted in many jurisdictions, is the preparation of an environmental management plan (EMP) to guide project implementation and operation. The contents and the detail of an EMP may vary enormously from one jurisdiction to another, but it seems to be generally understood that it contains tailor-made guidance to the implementation of all mitigation, compensation and enhancement measures. It may also contain – and indeed it should – for each measure or programme, a statement of objectives, definition of targets, timetables and milestones, as well as a procedure to evaluate actual outcomes.

Thus, an EMP 'translates' the mitigation and other measures emanating from the EIA into detailed instructions and actions to be implemented by the proponent during the construction, operation and decommissioning of its project. The extension to what such plans can be effective in delivering mitigation will be discussed in Section 2.3. For now, the point is that information and knowledge must flow downstream from the project assessment to the project implementation phases (Figure 2.1). Hence, the first question addressed in this chapter is 'How best can we ensure that information and knowledge generated in the pre-decision phase of EIA will flow down and ensure appropriate environmental protection outcomes?'

A number of practical difficulties may arise when 'translating'. Tinker et al. (2005) found that not all commitments described in a sample of EIS prepared in England ended up as legal conditions and obligations. In a sample of cases in Brazil, Dias and Sánchez (2000) found that terms and conditions of environmental licenses are often not directly enforceable and even less auditable, as they were written in vague and inaccurate language allowing for multiple interpretations. To overcome such hurdles and reduce the communication noise between assessment and implementation, some companies take care in preparing specific environmental guidance for construction or operation, such as a 'mitigation handbook' (Marshall, 2005). Some governments require a detailed transcription of all environmental commitments, such as the Hong Kong Environmental Protection Department (Hong Kong Environmental Protection Department, 1998) or the Western Australian Environmental Protection Authority.

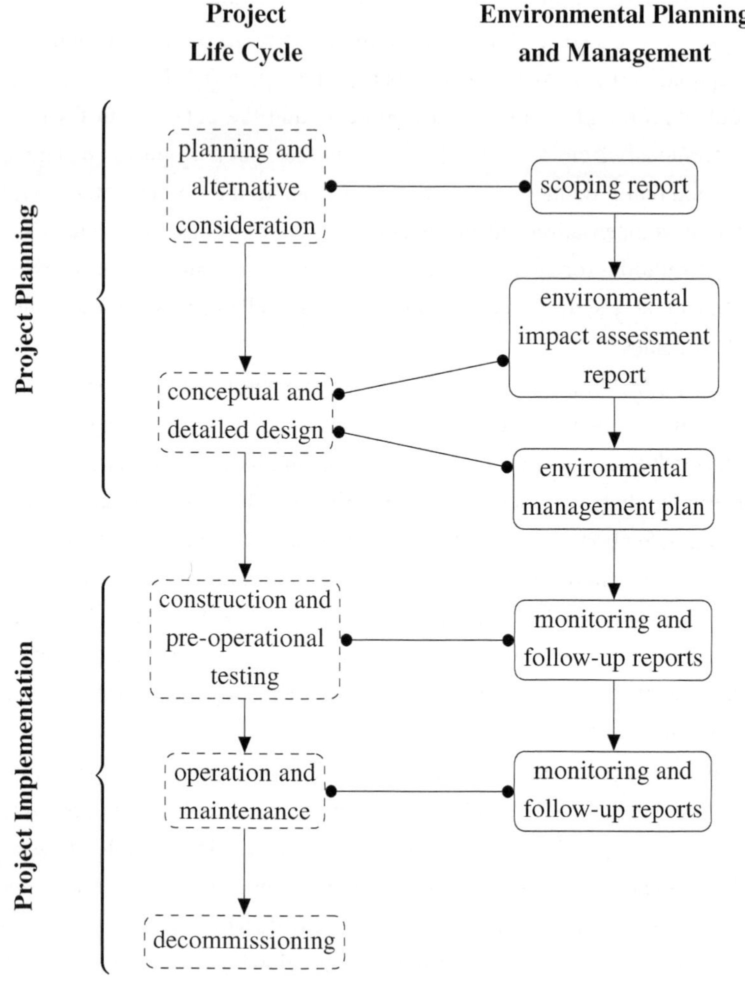

Figure 2.1 *The main steps in a project life cycle and the main documents generated in the EIA process; links are represented with horizontal lines*

It should not be assumed that because conditions are set by a government (or, for this matter, a financial institution) they will automatically be fulfilled. Deviances from expected can arise from a number of reasons, ranging from deliberate non-compliance or fraud to insufficient capacity or to accidents. In the remaining of this chapter, it will be assumed that proponents, contractors and consultants act in good faith to achieve environmental protection objectives agreed upon in the EIA process.

Management outcomes should be registered, documented and evaluated. Monitoring is required by law or regulation in many jurisdictions and indeed recommended as best practice. Monitoring can fulfil several functions, including: (1) checking the actual impacts of the undertaking; (2) detecting unpredicted impacts and warning against deviances from expected performance, thus triggering corrective action.

Besides these immediate and very practical functions, there is a huge potential for monitoring and follow-up to improve the environmental assessment of new projects. Not all jurisdictions call for comprehensive follow-up and reporting, but most require monitoring of actual impacts. Hence, the second question addressed in this chapter is 'How can information generated in the follow-up phase be transformed into knowledge which in turn can be useful for the management of ongoing operations and for the assessment of new projects?'

The first question will be dealt with in Section 2.3, the second in Section 2.4. Prior to this, in Section 2.2 the array of currently available tools will be summarized.

EIA is a knowledge-intensive activity. In preparing the documents outlined in Figure 2.1, the project proponent, the consultant, the government authority in charge of managing the EIA process (or 'decision making authority') and the public (hereafter called the main 'participants') generate data, compile information, use and create knowledge. Table 2.1 summarizes the relative role of knowledge detained by those participants according to the main steps in the EIA process.

Table 2.1 Relative part of knowledge holders in the main steps of the EIA process; legend: 0 = very low or no influence; + = low influence; ++ = important influence; +++ = very important influence

	Knowledge holders			
Step of the EIA process	**Consultant**	**Proponent**	**Public**	**DMA**[a]
Scoping/defining terms of reference	+++	++	++	+++
Baseline studies	+++	+	+	0
Impact analysis	+++	+	+	+
Design of mitigation and the EMP	+	+++	+	0
Decision making	0	+	+	+++
Follow-up	+	+++	+	+++

[a]Decision making authority

2.2 A Toolbox for Environmental Planning and Management

There is an ever increasing set of tools for environmental planning and management (EP&M) – Table 2.2. In this toolbox, managers have an array of instruments to choose according to their needs. Only large multinational companies are expected to use all or most of the tools currently available. The table is organised in three main types of tools, according to their functions: analytical, organisational and communicational. Tools were developed for different purposes, but many share common foundations and are built upon similar concepts. Dates of initial development of each tool are not shown, but EIA – having pioneered the structured and systematic consideration of the environmental consequences of today's decisions – is at the origin of many currently available tools. A number of analytical tools apply to projects whereas others apply to products or services, such as life cycle assessment, or the environmental assessment of products. This chapter deals only with projects and the organisations which conceive, approve, implement and operate those projects capable of causing significant environmental impact.

Table 2.2 Tools for environmental planning and management in organisations – modified and updated from Sánchez (2006); references cited are the most common and widely acknowledged at the international level; other pertinent references may apply

Tools	Guidelines and references
Analytical tools	
Environmental impact assessment[a]	IAIA Principles[b] World Bank safeguard policies[c] Convention of Biological Diversity Guidelines[d] IFC Performance Standards and Equator Principles[e]
Technological risk assessment	Seveso Directive[f] Guidelines AIChE[g]
Ecological and human health risk assessment	EPA[h] ASTM standards[i]
Environmental audit	ISO 19011:2002[j]

(Continued on next page)

[a]Includes social and health impact assessment, sometimes featured as separate tools.

[b]A series of leaflets depicting recommendations of best practice by the International Association for Impact Assessment (IAIA), available at www.iaia.org/publications/

[c]*World Bank Group Safeguard Policies*, available at www.worldbank.org/safeguardpolicies

[d]Decision VI/7, 6th Conference of Parts (The Hague, 2002) and Decision VIII/28, 8th. Conference of Parts (Curitiba, 2008); available at www.biodiv.org/convention/cops.asp

[e]*International Finance Corporation*, available at www.ifc.org/ifcext/enviro.nsf/Content/EnvSocStandards. Other tools include Guidance Notes and Environmental, Health and Safety Guidelines. The *Equator Principles* are subscribed by a number of financial institutions to deal with the social and environmental impacts of project financing, and are available at www.equator-principles.com

[f]European Union Directive on the control of major-accident hazards involving dangerous substances – see ec.europa.eu/environment/seveso/index.htm

[g]American Institute of Chemical Engineers – see www.aiche.org/Publications/

[h]Several US Environmental Protection Agency publications – see www.epa.gov/risk

[i]Supersedes ISO 14010, ISO 14011 and ISO 14012, by the International Organisation for Standardisation (ISO) – see www.iso.org

[j]Technical standards from the American Society for Testing and Materials, Committee E47 on Biological Effects and Environmental Fate – see www.astm.org/COMMIT/COMMITTEE/E47.htm

Tools	Guidelines and references
Environmental site assessment	ASTM E 1527-05 ASTM E 1528-06 ASTM 1903-11 ISO 14015:2001[k]
Environmental monitoring[l]	ISO 14004:2004 ISO 14031:1999
Environmental performance evaluation	ISO 14031:1999 ISO 14032:1999
Social accountability	SA 8000:2008[m]
Life cycle assessment	ISO 14040:2006 ISO 14044:2006 ISO 14047:2003 ISO 14048:2002 ISO 14049:2000
Greenhouse gas emissions	ISO 14064:2004 ISO 14065:2007
Organisational tools	
Environmental management systems	ISO 14001:2004 ISO 14004:2004 EMAS[n] Responsible Care® Codes of Management[o]
Workplace health and safety management systems	OHSAS 18001:2007[p]
Emergency response[q]	APELL[r]

(Continued on next page)

[k] Although there is an ISO standard, American site assessment standards are used internationally; other standards apply – see http://www.astm.org/COMMIT/SUBCOMMIT/E5002.htm

[l] Monitoring is essential to environmental management; however, there is no standard specific to monitoring.

[m] Social Accountability International, SA 8000:2008 – see www.sa-intl.org. SA 8000 is a certifiable standard.

[n] Eco-Management and Audit Scheme, established by the European Union – available at europa.eu.int/comm/environment/emas/

[o] Responsible Care® is an international initiative from the chemical industry. Their codes of management practice contain guidance specific to this sector.

[p] Occupational Health and Safety Standard.

[q] Emergency response is a part of management systems that requires interaction with the community. Community Awareness Emergency Response is a code from Responsible Care®, but the concept can be applied to any type of activity.

[r] Awareness and Preparedness for Emergencies at Local Level, promoted by the United Nations Environment Programme.

Tools	**Guidelines and references**
Corporate social responsibility	OECD Guidelines for multinational corporations[s] AA 1000:2008[t]
Environmental accounting	FASB standards (USA) and equivalent in other countries[u] UNCTAD/ISAR[v] Guidelines
Communicational tools	
Environmental performance reporting/sustainability reporting	GRI Guidelines – G3 (2006)[w]
Environmental labelling and certification	ISO 14020:2000 ISO 14021:1999 ISO 14024:1999 ISO 14025:2006 Sectoral certification programmes[x]
Business communication programmes	ISO 14063:2006

These tools can be used in a selective and integrative way to achieve environmental protection outcomes. Actual integration is hindered by a number of hurdles, ranging from excessive specialisation among practitioners (the EIA coordinator hands over to the EMS head who in turn, having consulted with the risk assessor, passes the ball to the compliance/performance evaluator who, then, informs the sustainability reporter) to fragmented bureaucracies – e.g. the branch in charge of follow-up is separated and staffed differently from the EIA team in the government Department of the Environment.

[s] Voluntary principles and standards for responsible business conduct in areas such as human rights, environment, information disclosure, combating bribery and consumer interests. Updated in 2011 and available at www.oecd.org/daf/investment/guidelines

[t] Three standards on sustainability assurance and stakeholder engagement. Available at http://www.accountability.org/standards/index.html

[u] FASB – Financial Accounting Standards Board, especially Statement of Financial Accounting Standards N° 143 on Accounting for Asset Retirement Obligations. Available at www.fasb.org/pdf/fas143.pdf

[v] ISAR – International Standards of Accounting and Reporting is a United Nations initiative that prepared guidelines for Environmental, Social and Governance Disclosure – see www.unctad.org

[w] GRI – Global Reporting Initiative, *Sustainability Reporting Guidelines 2006* (3rd version). Available at www.globalreporting.org/ReportingFramework/G3Guidelines/

[x] There are several certification programmes; a pioneer was the certification of forest products under the Forest Stewardship Council.

2.3 Integrating Project Planning and Management

In EP&M terms, the transition from project planning to implementation and then to operation and decommissioning calls for the integrated use of a number of tools listed in Table 2.2. Probably the most obvious integration is between EIA and an EMS. Among the advantages of such integration, efficiency gains are widely recognised, as it provides room for cost reduction by eliminating duplication of efforts (Ridgway, 1999; Sánchez and Hacking, 2002; WBCSD, 1996). Varnäs et al. (2009) suggest that green procurement can be successfully integrated into the EIA–EMS nexus and provide a case study. Evidence of successful EMS in support of compliance with terms and conditions originated in the EIA process are also provided by Sánchez and Gallardo (2005) for a highway and by Varnäs et al. (2009) for a railway. There are probably several success cases, but only a few are documented in the academic literature. On the other hand, failure to achieve such an integration has also been documented. Slinn et al. (2007), after identifying the poor EIS-EMS linkage in the planning of new industrial estates in the UK, consider that having the decision making authority requiring the preparation and implementation of an EMP could be an effective way to make EIA a facilitator of environmental management.

Another advantage of EIA–EMS integration is that it allows for focusing EIA on management capacity and not exclusively on prediction accuracy (Goodland and Mercier, 1999; Noble and Storey, 2005). Ridgway (2005, p. 325) states that it is becoming clear that EMS 'can be used to assist the delivery of EIA outcomes'.

A third advantage of integrating EIA, EMS and other tools is that it enables organisational learning. This potential has so far been overlooked in the EIA literature, although practitioners seem to capitalize on this fact – any environmental consultancy showcases its list of projects and clients, implying the experiential knowledge it gained from previous undertakings. It can be argued, it follows, that organisations with multiple projects and several past and ongoing EIAs may learn from experience and selectively incorporate learning into future assessments.

The transition from project planning to implementation and operation can be facilitated integrating EIA and EMS and other management tools. Sánchez and Hacking (2002) propose that an EMP should contain not only 'traditional' mitigation measures, but also 'capacity' management, i.e. human resources and management. Goodland and Mercier (1999) strongly emphasise that successfully implementing an EMP requires (i) securing a budget, (ii) defining an implementation schedule, and (iii) providing adequate human capacity.

Ridgway (2005) recommends building a register of commitments stated in the EIA, as regulations in many jurisdictions do not call for a consolidated list of all management measures that may be scattered in EIA documents. Each commitment could then be incorporated into the EMS as a document that needs to be kept up to date.

In short, both the literature and practical experience show that the transition from planning to management can be achieved through a convenient mix of managerial tools and capacity development.

2.4 Managing the Project for Achieving Environmental Outcomes

Management can be conducted on the basis of programmes or systems. An EMP is typically made up of several programmes (e.g. water management, land rehabilitation, waste management etc.). Each programme can be run independently, with its own objectives, targets, time schedule, budget and team. Alternatively, the programmes can be integrated in a system – i.e. an environmental management system or an integrated management system encompassing environment, health and safety and quality. The advantages of systems have been advanced and reviewed in the literature (Tinsley and Pillai, 2006) and include not only possible better environmental performance, but also better overall performance for firms that adopted and certified an EMS (Melnyk et al., 2003).

In EIA terms, management is part of the post-approval or follow-up phase, defined as 'the monitoring and the evaluation of the impacts of a project or plan (that has been subject to EIA) for management of, and communication about, the environmental performance of that project or plan' (Morrison-Saunders and Arts, 2004a, p. 4).

According to Arts and Morrison-Saunders (2004, p. 25), the need for follow-up arises from the uncertainties inherent to any prospective exercise. Without follow-up, the 'outcomes of EIA decisions and actions cannot be determined and *learning from experience* cannot occur' (emphasis in the original).

Follow-up requires a shared participation of both project proponent and the regulator, with a role for stakeholders. Tools for follow-up are: monitoring, inspection, government overseeing, supervision, auditing, register and performance evaluation. The meaning of these terms is not universal and they are sometimes used interchangeably or with different interpretations according to the speaker. Working definitions are provided:

- monitoring is the systematic and periodic collection of selected data;
- inspection is 'an examination into the environmental affairs of a single regulated facility' (US EPA, 1989, pp.3–8); on-site inspections activities may include records review, interviews, physical sampling and observations (US EPA, 1992); although the term is usually employed to describe the overseeing activities of a government agent, it is sometimes also employed to mean an overseeing activity undertaken by any agent, such as a representative of a financial institution, like the 'inspection panels' of the World Bank;
- supervision is a continuous activity undertook by the proponent or on his behalf aiming at checking legal or contractual compliance by contractors involved in construction or operation activities; environmental supervision is being used in civil works as a means of ensuring that terms and conditions of licenses and authorizations are met;
- auditing, as defined in the context of management systems, is a systematic, independent and documented process for obtaining audit evidence and evaluating it objectively to determine the extent to which the audit criteria are fulfilled (ISO, 2009);
- register is the systematic storage of data and information such as monitoring results, nonconformities documentation, compliance evidences, lists of environmental aspects and impacts;
- environmental performance are the 'results of an organisation's management of its environmental aspects' (ISO, 1999) or, more precisely, the 'measurable results of an organisorganisation's management of its environmental aspects' (ISO, 2009); environmental performance evaluation is the 'process to facilitate management decisions regarding an organisation's environmental performance by selecting indicators, collecting and

analysing data, assessing information against environmental performance criteria, reporting and communicating, and periodically reviewing and improving the process' (ISO, 1999).

Figure 2.2 shows the use of tools during the post-decision phase. The respective roles of proponent, government and the financial institutions in the follow-up phase are supported by different instruments.

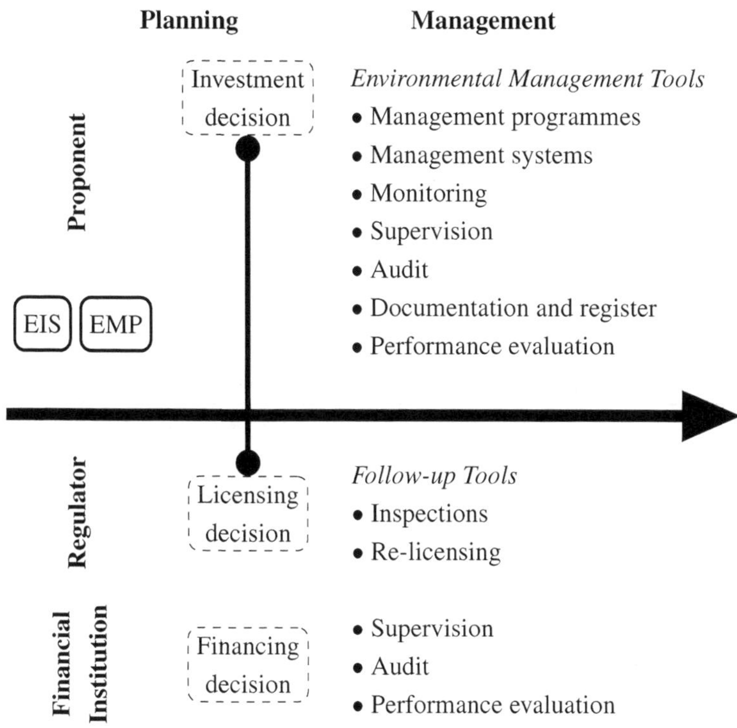

Figure 2.2 Management tools for EIA follow-up

Follow-up reports are usually required to be filed with the relevant government authority (Baker, 2004). Those reports and the follow-up activities generate data, information and knowledge which can be used for several purposes, including:

- compliance assessment
- conformance assessment
- triggering corrective action or adaptive management
- recording results
- performance evaluation
- assessing effectiveness and efficiency of mitigation
- environmental or sustainability reporting

The meaning of the terms data, information and knowledge is often dubious not only in plain language, but also in the literature. For the purposes of this discussion, the practical definitions are shown on Table 2.3, adapted from Liew (2007), who reviewed definitions and concepts proposed by several authors, and an expert group set up under the banner of the European Committee for Standardization (CEN, 2004).

Table 2.3 Data, information and knowledge in EA&M; modified from CEN (2004) and Liew (2007)

Term	Concept	Examples
Data	Captured (recorded and stored) symbols and signal readings Objective facts (numbers, symbols, figures) without context and interpretation	Raw results from monitoring Water consumption at an industrial plant
Information	A message that contains relevant meaning for decision or action Data in context	Water quality at a certain location and moment Percentage of time emissions are below or above a threshold
Knowledge	Cognition (know-what) Acting capacity (know-how) Understanding (know-why, know-where)	Mitigation effectiveness Project's area of influence How to avoid or reduce impacts

Hence, data must be organised (i.e. selected, captured, categorised, indexed, stored and retrieved) to become information which in turn can be processed by people to make sense of situations or events, thus creating knowledge. Any follow-up report (in whatever form it may be presented to a government authority) features recorded data and information. Typically, a government agency does not want to receive files, spreadsheets or databases, but information about compliance, achievements or corrective actions. For proponents, follow-up results can be used to demonstrate compliance and conformance to regulators, investors, the community and other interested parties. Results can also be used internally for controlling on an ongoing basis, such as in adaptive management.

The key point is that making sense of follow-up results (data and information) represents a learning opportunity for project proponents. The extent to which this potential is realised depends on a number of reasons whose in-depth discussion is beyond the scope of this chapter, but a brief discussion will follow below.

After identifying barriers to learning and adaptation in natural resource management, such as organisational politics, culture and capacity, Genskow and Wood (2011) propose that overcoming them involve elements of information, structure and culture. Information, in this context, relates to adopting appropriate metrics to evaluate results, which is what monitoring, supervision and auditing are about. Structure relates to providing organisational 'learning environments' that allow, among several purposes, for shared interpretation of data and information – like establishing routines for data analysis by a diverse set of personnel (e.g. from the 'environment' and the 'production' departments). Culture relates to the elements of organisational culture that influence a learning environment, like a clear commitment to improvement from the leadership, which is also an essential element of ISO 14001. Hence, to enable learning from follow-up, a 'learning-organisation approach' (Bond et al., 2010) is a requirement.

The use of data and information obtained in follow-up activities for influencing proponent's behaviour in order to improve project environmental management is called here the 'first feedback loop of EA&M'. As it can be implied from above, there is no 'spontaneous' or 'automatic' feedback. It needs to be built into the organisation and an EMS provides an adequate framework as it is designed to foster continuous improvement.

Additionally, it has been advanced that follow-up can also play a role in improving future environmental assessments – Arts and Morrison-Saunders (2004) among others. Thus, if a proponent has several ongoing and future projects, it can benefit from follow-up for:

- deriving standard mitigation for future projects;
- establishing or updating management procedures;
- improving criteria for assessing the importance of impacts;
- improving criteria for assessing outcomes and performance;
- reducing uncertainty for future assessments.

Large companies that continuously develop new projects certainly can benefit from establishing internal routines and procedures and replicating best practices adopted in earlier projects and their respective environmental assessments. Such learning that outgrows individual projects derive from a 'second feedback loop of EA&M', i.e., experiential knowledge gained in one project feeding other projects. A few large companies may develop their own standards, such as the mining company Anglo American with its 'Socio Economic Assessment Toolbox', publicly available at the company's website. Of course, small companies that only occasionally undergo the EIA process, although can benefit from the first loop, could hardly benefit from the second. In this case, there is a role for environmental consultants to act in knowledge transfer.

Figure 2.3 synthesizes the relationship between selected EP&M tools and the two feedback loops. The bold lines from EMS to monitoring and audit and back to EMS correspond to the first loop, while the dashed line from environmental performance evaluation to future environmental impact assessments represents the second loop; the circuit from EMS to performance evaluation and back to EMS is also part of this second loop.

2.5 Managing the Process for Achieving Environmental Outcomes

The fact that there is a learning potential does not mean that organisational learning will occur. The literature has shown that organisational learning depends on a number of factors, including organisational culture and leadership (Argote et al., 2003; Genskow and Wood, 2011). As a matter of fact, there are several cases where the learning potential provided for follow-up is far from being realised. In Brazil, an audit conducted by the Federal Accounts Tribunal found that the National Department of Transportation Infrastructure attained positive environmental protection outcomes in a few highway construction sites, but

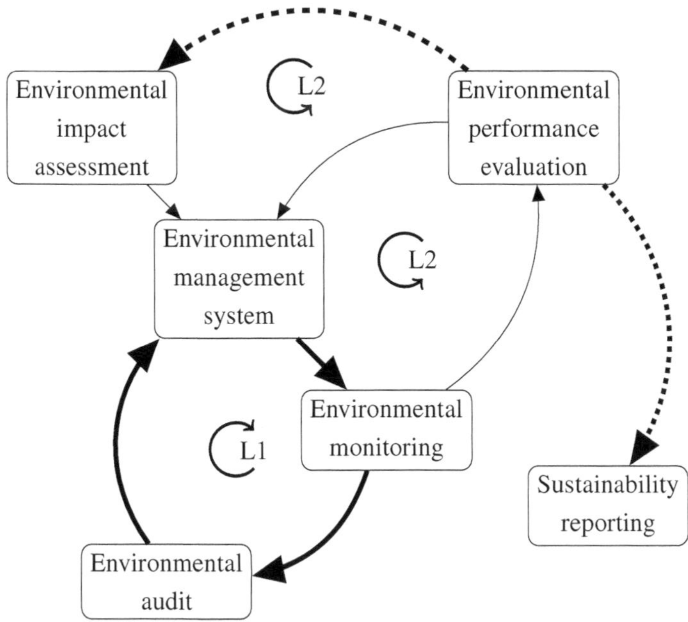

Figure 2.3 L1 and L2 in environmental assessment and management

failed to transfer knowledge to new projects (the second feedback loop). One reason for failure is that this Department outsourced most follow-up activities, thus did not retain knowledge and could not make it available for new projects.

In contrast, environmental consultants working for one São Paulo State highway agency perceived noticeable improvements when comparing project planning practices over a 12-year time span, between the first and the fourth and last reach of the São Paulo Metropolitan Ringroad. In this agency, although engineering projects and environmental assessments are commissioned from specialised firms, the agency management mediates negotiations between both teams to find acceptable compromises between technical and economic project requirements and environmental protection needs, building on the previous experience of both external teams and the agency's staff.[1]

[1]Opinion collected through interviews in the context of an ongoing research whose results have not yet been published.

That management consultancy firms may build experiential knowledge and make use of it for business purposes is well known and explored in the literature (Werr and Stjernberg, 2003). The same is intuitively true for environmental consultants involved in EA&M. Bond et al. (2010), discussing knowledge validation within an EIA consultant team observe that, although both formal (explicit) and informal (implicit) knowledge are used in EIA, informal procedures for knowledge acquisition and validation play a central role. They then ponder that a learning-organisation approach is needed if EIA is to achieve its aim of fostering sustainable development.

Finally, it is not only the proponent or the consultant that can learn from experience by exploring the results of follow-up. Both the government agencies in charge of managing the EIA process and the financial institutions funding projects can derive organisational learning from environmental follow-up and possibly apply such knowledge to improve procedures or to achieve better outcomes.

Although EIA has been evolving in several jurisdictions, and its perceived effectiveness is considered when reforming laws or institutions, there seems to be no structured approach by EIA agencies to systematically harness knowledge created through follow-up. Sánchez and Morrison-Saunders (2011) found very limited use of information generated in the follow-up phase in Western Australia to improve practices of the State EIA agency, in spite of this agency's initiatives to capture, store and disseminate experiential knowledge.

On the other hand, Buntaine (2011), who reviewed the lending practices of the Asian Development Bank, collected empirical evidence to support his finding that the institution responds positively to previous borrower environmental performance when making decisions of approving new 'environmentally risky projects'. Caspary (2009, p. 27), reviewing the EIA procedures adopted by public financial institutions for 20 dams in different countries, found that multilateral development banks (differently from bilateral export credit agencies) tend to develop 'regular ex-post evaluation of environmental impacts of project portfolio'.

That follow-up can be beneficial for improving the management of the EIA process in a particular jurisdiction or organisation (like a financial institution) can be described as the 'third feedback loop of EA&M'. From the first to the third feedback loops there is increased complexity, larger timeframes and, generally speaking, more difficult barriers to be overcome. Many proponents may simply not perceive to be in their best interest the sharing of knowledge derived from follow-up. Many governments, on the other hand, may not offer the political support and the resources needed to enable the third loop, as they are under pressure from short-term business and political interests that see the EIA process as cumbersome and time-consuming. Community groups and a number of civil society organisations, on the other hand, often do not – or cannot – act beyond the limits of a single project they may oppose or wish to influence, thus are not in a position to influence an EIA system.

Arguably, advancing practices by benefiting from a third feedback loop of EA&M involves some degree of social learning in addition to organisational learning (Saarikoski, 2000; Sinclair et al., 2008), but this is a matter to be explored in further detail.

2.6 Synthesis and Conclusions

Two questions were proposed for discussion in this chapter. First, it was asked 'How best can we ensure that information and knowledge generated in the pre-decision phase of EIA will flow down and ensure appropriate environmental protection outcomes?'

Both theoretical literature and empirical evidence obtained from case studies and practitioners' experience suggest that robust environmental management plans stocked with clearly stated objectives, goals, budgets, time schedules, designation of responsibilities and staffed with competent personnel can ensure a successful transition from environmental planning to management, i.e. from paperwork to work on the terrain. The best EMPs prepared today reflect, in first place, our collective learning about issues such as effectiveness of mitigation measures and how to design effective mitigation (e.g. how to plan and build a fauna passage in a highway in a particular environment), and second, learning achieved by the project proponent and its consultants.

The second question addressed in this chapter is 'How can information generated in the follow-up phase be transformed into knowledge which in turn can be useful for the management of ongoing operations and for the assessment of new projects?' Making sense of information generated in follow-up for controlling and influencing a project's management makes up the 'first feedback loop of EA&M'. Making sense of information and knowledge derived from follow-up to improve the assessment and the design of mitigation for future projects constitutes a 'second feedback loop of EA&M'.

Both loops can be detected, at varying intensity, in several organisations – proponents, consultants, government agencies and financial institutions. There is both organisational and social learning in the EIA process. However, learning opportunities can easily be missed or underexplored unless a 'learning organisation approach' is adopted by those players in the EIA process, as simply filing monitoring reports by no means leads to learning. Fostering a learning climate is a major role that only the government agencies in charge of managing the EIA process can play.

3. Link Framework Analysis

Anastássios Perdicoúlis

The way we currently deal with environmental impacts of development or investment projects is somewhat fragmented. At the study phase, or *ex-ante*, the predominant instrument is environmental impact assessment (EIA). After implementation, or *ex-post*, the predominant instrument becomes environmental management systems (EMS). While both instruments are established with well known procedures, their integration remains a significant challenge.

It is difficult to argue that the *ex-ante* and *ex-post* procedures should remain independent, since they are both dedicated to the same object – that is, environmental impacts of development or investment projects. If we argue that the two instruments have different perspectives – for instance, regarding time – then they should be at least complementary to each other. If we argue that EIA and EMS take different perspectives on their common object of interest – for instance, environmental accounting versus financial accounting – then they should be taken together, at least to juxtapose their respective perspectives.

In just over a decade, the attempt to link EIA and EMS has been manifested in a number of publications – for instance, Barnes and Lemon (1999); Ridgway (1999); Eccleston and Smythe (2002); Sánchez and Hacking (2002); Perdicoúlis and Durning (2007). The existing attempts contain various degrees of formality and strength of the link, and collectively elevate the issue to a higher level: What could be the *framework*, or general method, to establish the link between EIA and EMS?

There are many ways to search for link frameworks, as illustrated in Figure 3.1. Of the alternative approaches, analysis methodically examines and elucidates processes and/or functions, whereas design is guided by general interests or concrete objectives. Finally, simile seeks other known examples of link frameworks, in similar or unrelated fields, and brings in their experience – or merely copy what others do. Although it is possible to combine these approaches, let us attempt something easy and realistic to explore some options of link frameworks for EIA and EMS in this chapter.

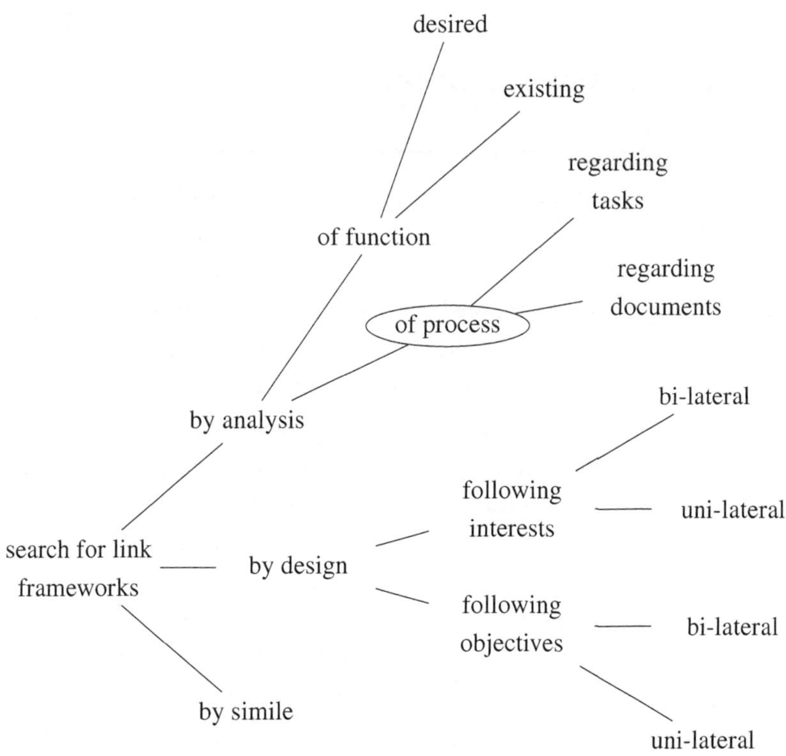

Figure 3.1 There are many ways to search for link frameworks; the chosen one for here is 'process analysis'

Since there already exist a number of EIA–EMS link attempts, it seems like an obligation to consider them in an analytic manner. This analysis can be done regarding the function of the link – for instance, what the link should facilitate – which is very abstract, or alternatively regarding the process of the link, which is more concrete and focused. Therefore, and considering the limited space of the chapter, the chosen approach for the EIA–EMS link framework is the 'process analysis' method.

3.1 Analysis

3.1.1 Plan

Let us implement the process analysis through two 'filtered views': first to see only the tasks of the two instruments, and search how these could link, and second to see only their documents and how these could link. Finally, in an integrating view, we can merge the two views plus what we learned in each one, to obtain a realistic 'composite' view of possible frameworks – Figure 3.2. Throughout this analysis we shall consider existing work, and also explore new possibilities pointing to future research.

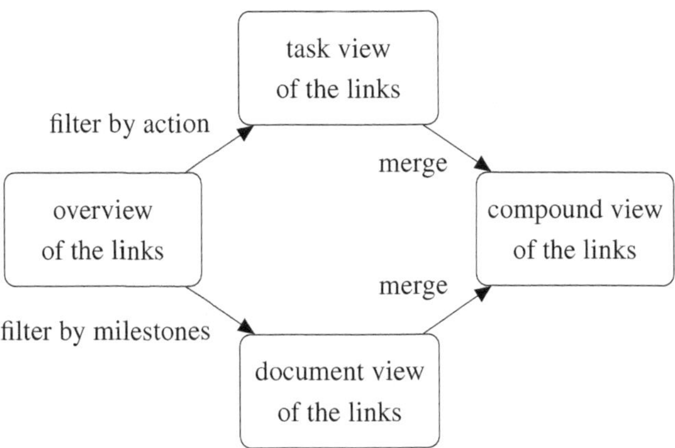

Figure 3.2 Through process analysis, the chapter features four alternative views of the links between the two environmental instruments, EIA and EMS

The analysis presented in this chapter is quite general or theoretical, which is suitable to obtain wide perspectives and high-level lessons. This may need to be accompanied by further detailed essays involving specific case studies. Although the issue may seem relatively straightforward, and it is definitely a pressing issue, the time frame for this analysis and further developments is not likely to be short. After theoretical analyses, concepts must be proven in practice, and this requires time, good will, effort, and institutional support.

3.1.2 Process

To start with the process analysis, let us consider the concept of process and ways to represent it. As a succinct definition, a process is a series of actions (or tasks), and a number of corresponding stages (or steps) which are shaped by the actions. Some tasks can be carried out in parallel, thus leading to complex (or multi-task) processes. In addition, there may be delays between the tasks and the following stages, or the stages may not be perfectly perceivable, thus leading to complicated processes characterised by uncertainty (Perdicoúlis, 2011).

To make this analysis more organised, and thus easier to understand, let us briefly work out the relevant concepts and nomenclature. First, the stages of a process (also known as 'steps'), often take the form of input or output documents – for instance, the volume of data required for a project, or the final report of the project. Groups of steps – that is, at a higher aggregation tier – are known as 'phases'. The action of a process, which is often required between its stages (also known as 'tasks') is typically expressed in action verbs – for instance, collect the data, or type the project report (Perdicoúlis, 2010).

3.1.3 Perspectives

We can visualise and communicate processes in two different perspectives, depending whether we are more interested in the action or in the stages. In the 'action perspective' of a process (Figure 3.3), the tasks are the most important elements; the stages of the process may even be optional elements, in which case the process is reduced to a task sequence.

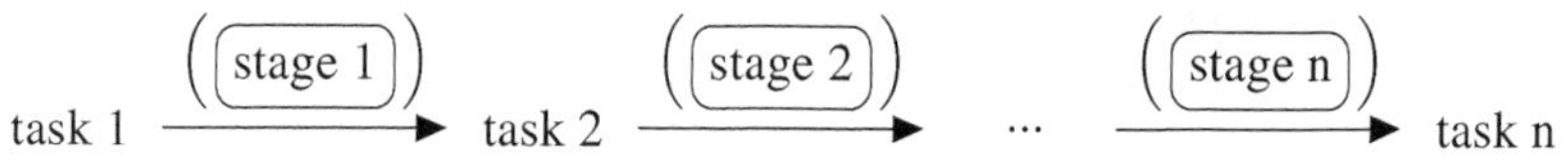

task 1 —(stage 1)→ task 2 —(stage 2)→ ... —(stage n)→ task n

Figure 3.3 A single-track process in an 'action perspective': the tasks are the main elements, while stages are optional

On the contrary, in the 'milestone perspective' of a process (Figure 3.4), the main elements are the stages. The action may even be optional, in which case the process is reduced to a sequence of stages.

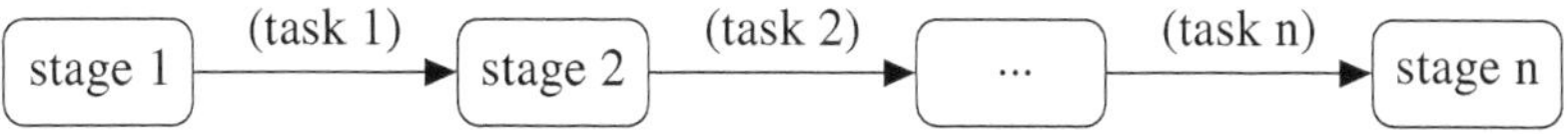

Figure 3.4 A single-track process in a 'milestone perspective'; stages are the main elements, while tasks are optional

Besides these partial perspectives, we may be interested in both the action and the stages, in which case our perception and representation are more realistic and complete. To facilitate the analysis, we shall start the chapter with the partial perspectives before attempting the composite view.

3.2 Overview

3.2.1 Timing Reference

Regarding their localisation in relation to the implementation of the project, the two instruments have an apparent – and potentially conditioning – link order when considered as closed or generic blocks. A distant or aggregated look at EIA and EMS in this perspective may be represented in Figure 3.5.

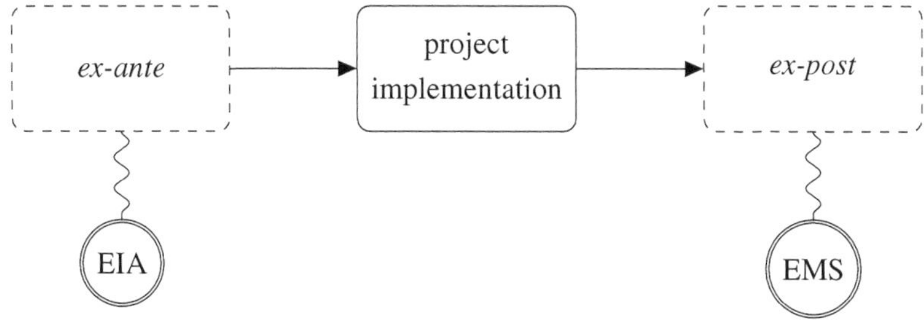

Figure 3.5 Overview of the links between the two environmental instruments, EIA and EMS, in relation to the project timeline

But although EIA has been conceived as an *ex-ante* environmental instrument, justified to exist before the implementation of the project, it seems to be constantly extending beyond the implementation of the project – for instance, the importance and popularity of EIA's 'follow-up' phase has been increasing over the last decade (Morrison-Saunders and Arts, 2004a). In a similar manner, EMS has been conceived as an *ex-post* instrument, justified to exist only after there is a project, but projects are not meant to appear spontaneously: they are expected to be guided by existing policies – for instance, regarding development or investment, and the environment. These two extensions may be blurring the clear boundaries of Figure 3.5, but they are promising some 'sticky ends' for the link of EIA and EMS.

3.2.2 Hindering Issues

Despite the promising indication that both EIA and EMS have been extending to each other's fields of jurisdiction, with reference to the crucial mark of project implementation, some hindering issues have been identified in a number of publications – for instance, as summarised in Perdicoúlis and Durning (2007) and illustrated in Figure 3.6 – and they stand in line to be addressed.

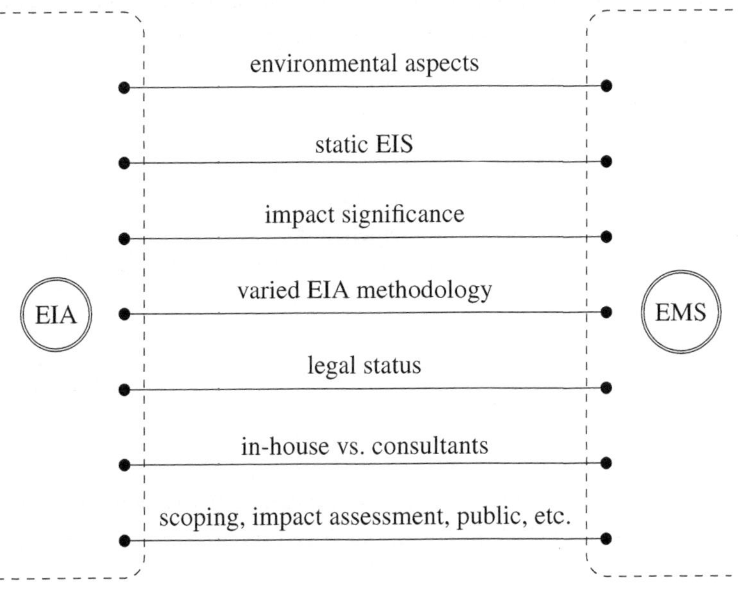

Figure 3.6 Hindering issues regarding the link between EIA and EMS

3.3 Action or Task View

Among the limited bibliography dealing with the EIA–EMS link, the typical view is about the tasks of the respective processes. This is understandable when the major concern is action, which is important as people know that they can be held responsible for their actions. Let us then consider analytically what the action view reveals for each process.

3.3.1 EMS

The standard EMS phases – for instance, as given by ISO (2004), and also followed in the EIA–EMS link attempts by Sánchez and Hacking (2002) – typically order the tasks in an EMS cycle as in Figure 3.7. Since this process is defined by ISO, this is naturally the reference pattern for international practice.

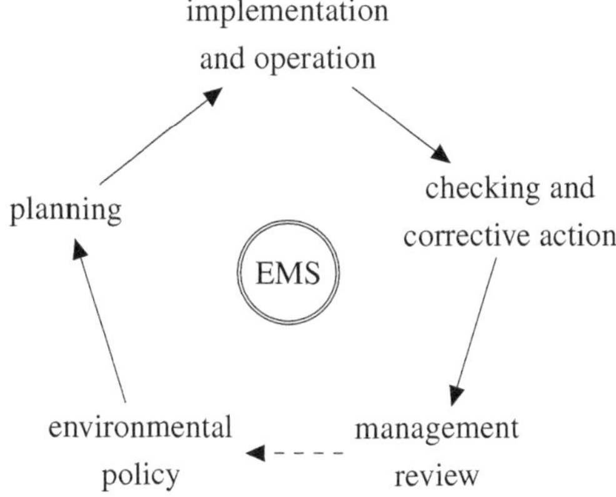

Figure 3.7 EMS phases; while 'policy' is a document, the corresponding action is intended as 'policy making'

3.3.2 Project

Project phases are expected to vary according to project management styles, but we can generally accept some organisation with a preparation phase, an implementation phase, the main operation phase, and possibly a decommissioning phase. In the EIA–EMS attempts, Ridgway (1999) has made suggestions for the attribution of appropriate environmental impact management instruments – that is, EIA and EMS – to those generic project phases, as illustrated in Figure 3.8.

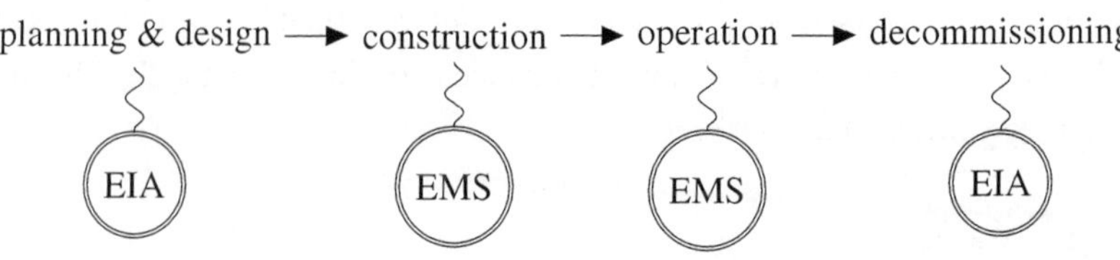

Figure 3.8 Generic project phases and suggestions for the attribution of appropriate instruments – after Ridgway (1999)

3.3.3 EIA

Similar to the project process, the EIA process is also subject to implementation styles. Again, we can distinguish generic EIA phases, as illustrated in Figure 3.9. We must note a recursiveness in the nomenclature of EIA, as 'assessment' refers both to the whole process – as in 'environmental impact assessment' – and also to a particular phase, where the forecast impacts are estimated and evaluated.

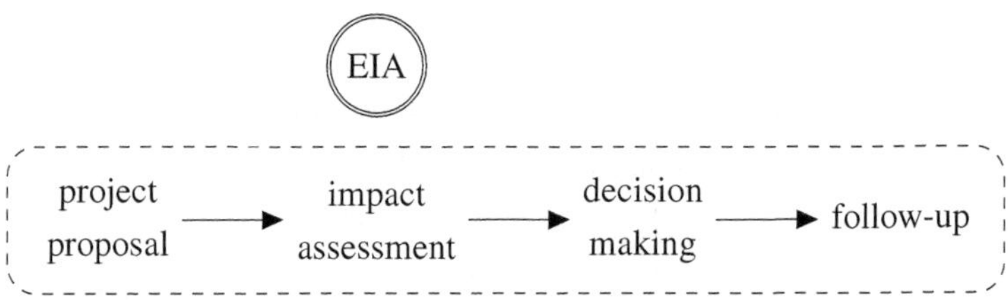

Figure 3.9 The EIA process features a 'speciality phase' of impact assessment, which is flanked by a number of ancillary phases

3.3.4 EIA and EMS

Some of the attempts to link the EIA and EMS processes have resulted in a number of configurations. Let us briefly present some of them, under a common graphical style that facilitates comparisons. To start with a relatively simple link suggestion, Sánchez and Hacking (2002), for instance, attribute the initial part of the EMS cycle to EIA, as illustrated in Figure 3.10. It is interesting to notice that the tasks forming the base of this sequence are the same of the EMS cycle, presented in Figure 3.7.

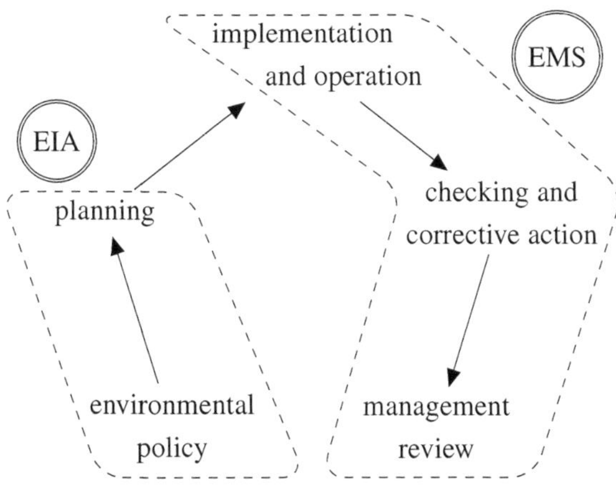

Figure 3.10 The initial part of the EMS cycle is attributed to EIA – after Sánchez and Hacking (2002)

In a more complex link solution, Eccleston and Smythe (2002) augmented the EMS cycle with EIA tasks, and re-distributed the responsibilities between the two processes – Figure 3.11.

Finally, and somewhat curiously, Perdicoúlis and Durning (2007) reach a more conventional distribution of responsibilities of tasks across the two instruments, as illustrated in Figure 3.12, but through a more complex alternating arrangement. In this case, 'conventional' indicates that the tasks are 'native' to their original processes – for instance, environmental policy (making) and planning belong again to EMS, and not to EIA as suggested by Eccleston and Smythe (2002) – compare Figures 3.11 and 3.12.

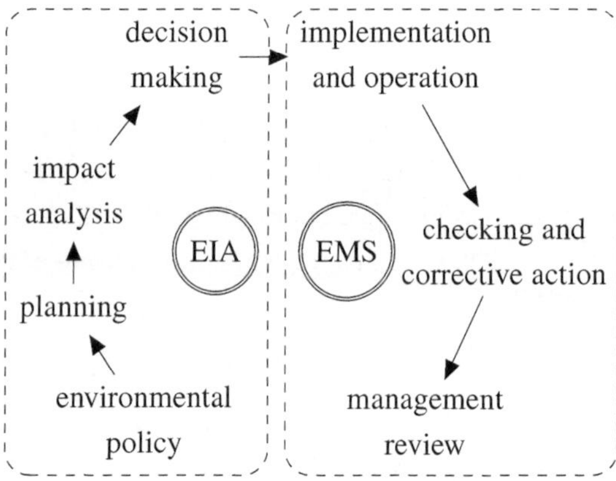

Figure 3.11 *An EMS cycle augmented with EIA tasks receives a re-distribution of responsibilities – after Eccleston and Smythe (2002)*

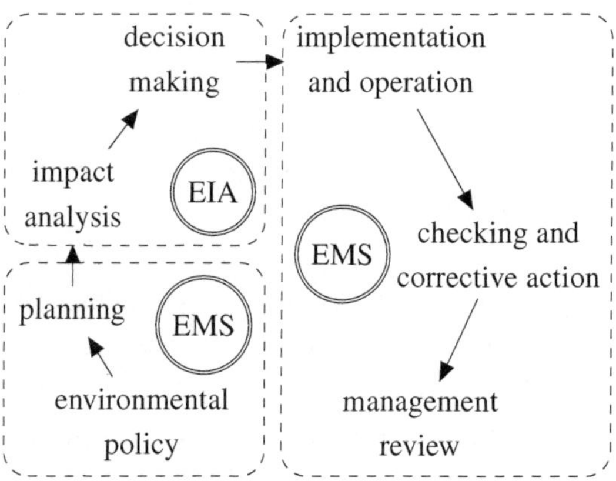

Figure 3.12 *The augmented EMS cycle of Figure 3.11 with a more conventional distribution of responsibilities – after Perdicoúlis and Durning (2007)*

3.4 Document View

Contrary to the mainstream action or task view of the EIA and EMS processes, a document view can be found in Perdicoúlis and Durning (2007), illustrated in Figure 3.13. This unusual 'document cycle' identifies crucial 'interfaces' in the EIA and EMS processes, which can help separate process boundaries or jurisdictions, and also indicate a sequence in the processing of the information contained in the documents.

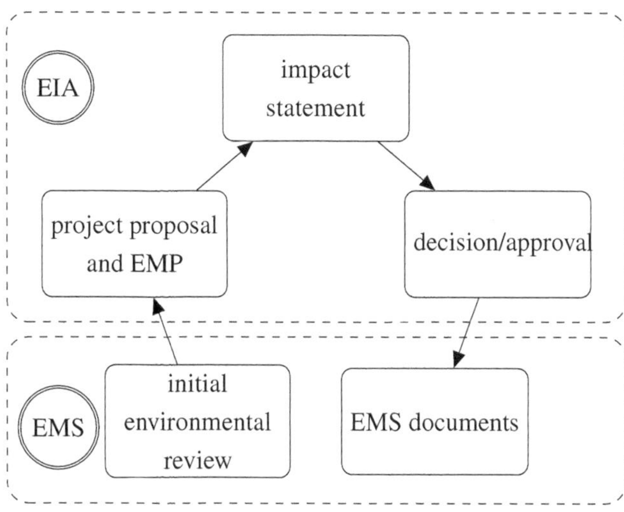

Figure 3.13 A document cycle identifies crucial 'interfaces' in the EIA and EMS processes – after Perdicoúlis and Durning (2007)

It is evident that Figure 3.13 is highly aggregated, and the production of each of these documents implies complex tasks. Although this aggregated approach is taken to match the scope of the rest of the material in this chapter, it is possible to add more detail to the document cycle – for instance, for further analysis or implementation purposes – by detailing each one of the boxes. As an example, the impact statement may contain an impact forecast volume, an impact assessment volume, etc. Finally, the environmental management programme (EMP) may appear as a novelty to classic EIA, but this is rapidly becoming common practice in EIA, as evidenced in other chapters of the book.

3.5 Integrated Process View

After the task-only and document-only analytic views of the EIA and EMS processes, separately and together, we can now take an integrated view on the processes. Although we are merging two previous views, this is still an analytic step. The difference is that the integrated view includes more complexity than the previous filtered views.

Since the integrated process view considers tasks and documents simultaneously, we have two options to represent the process: either as a sequence of tasks, or as a sequence of documents – that is, we have the option to to give priority to either tasks or documents in the description of processes. As we saw earlier in the chapter, this provides two analytic perspectives: the action perspective (Figure 3.14) and the milestone perspective (Figure 3.15).

3.5.1 Action Perspective

Figure 3.14 represents the integrated process view as a sequence of tasks – which are the plain text nodes – with the associated documents marked as rounded rectangles on the transition arrows between the tasks. The tasks, or nodes of the figure, are based on the 'enriched' EMS cycle, presented in Figures 3.11 and 3.12. The documents are based on the the document cycle of Figure 3.13.

This integrated action perspective (Figure 3.14) reveals two loops: the lower loop represents the EMS process, and the top loop represents the EIA process. The loops are being presented in an apparently modular arrangement – that is, each one is a recognisable process – but linked as one on top of the other. This modularity is important for an effective and easy link between the two processes.

Besides the modular arrangement of Figure 3.14, which was based on the particular versions of the 'enriched' EMS cycle and the document cycle, it is possible to create more advanced action perspective arrangements – not represented here. Laboriously innovative models may bring about valuable insights, but may face practical difficulties in the actual link of the two processes – at least as we know them.

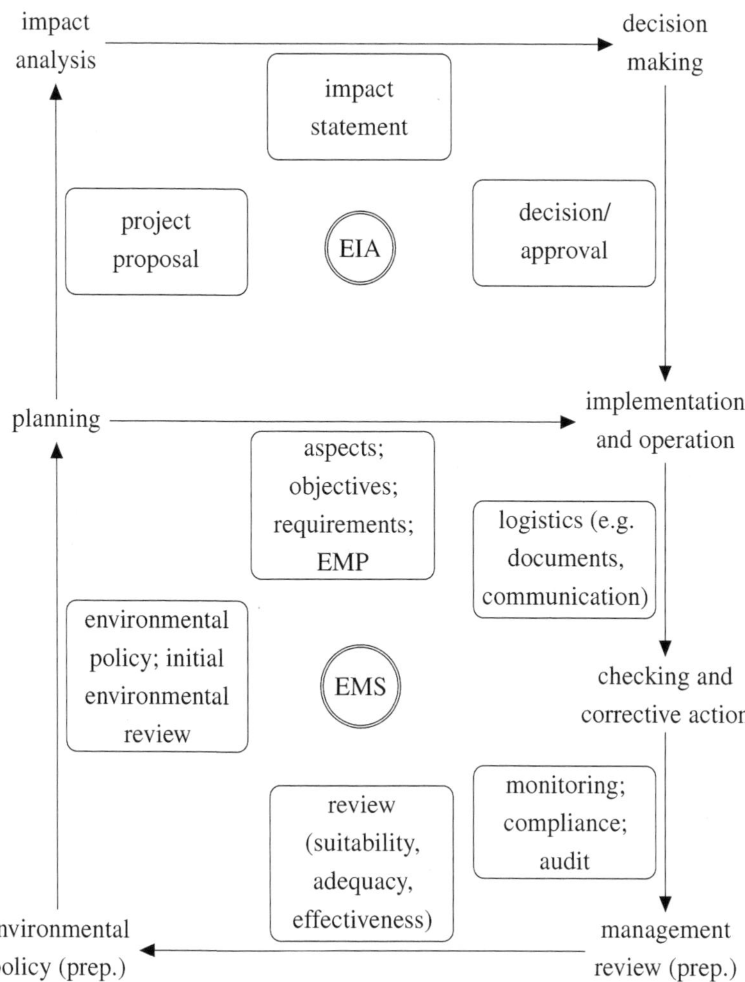

Figure 3.14 An action perspective of an 'enriched' EMS cycle, combined with the elements of the document cycle

3.5.2 Milestone Perspective

Figure 3.15 represents the integrated process view as a sequence of documents at the nodes, and the associated tasks on the transition arrows between the nodes. The documents, or the nodes of the figure, are a more elaborated version of the document cycle of Figure 3.13.

Although the action and milestone perspectives are inverse regarding the arrangement of tasks and stages, the latter has a special property – namely, a 'tangible' core sequence, composed of documents, as opposed to the more abstract action. This more 'tangible' node structure at the working base is capable of presenting more 'gravitas' or 'presence', which can then permit more freedom or flexibility in the definition or revision of the more abstract (or 'softer') elements – in this case, the tasks or action, marked on the arrows.

Figure 3.15 takes advantage of this 'tangibility' property of the milestone perspective to explore some innovation options – namely, when the tasks can possibly deviate from the original action scheme of the EMS cycle (Figure 3.7). In this case, the tasks presented in Figure 3.15 are a variation of those of the EMS cycle, with an opportunity to give more detail – for instance, 'information reporting' or 'position assessment'.

Once again we note that while the deviation from the standards or current practice may be considered as freedom to bring about more innovation, it may also challenge the existing EMS norms and EIA legislation, and can even make the EIA–EMS link more laborious.

Some examples of innovation taking advantage of the 'tangible' property of the milestone perspective are illustrated in Figure 3.15. For instance, the project proposal as the precursor of both the environmental (impact) statement, or E(I)S, and the nameless package that contains the 'planning' information of the EMS – namely, environmental aspects, objectives, and the environmental management programme (EMP).

Another interesting idea presented in the particular configuration of the milestone perspective, for instance, is the possibility to verify and adjust the 'planning document' (including EMP, aspects, and objectives) with the environmental policy or principles. Such an internal check may imply delays – often seen as 'bureaucracy' – but essential to maintain integrity in the content of the process.

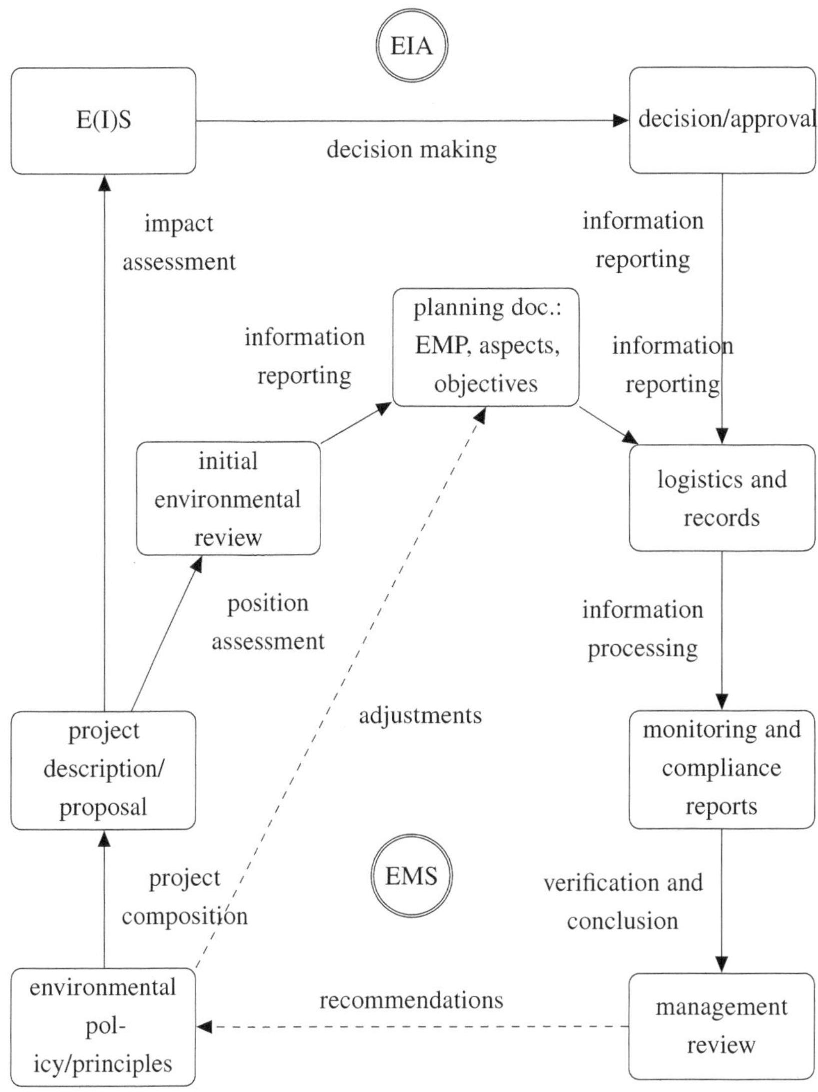

Figure 3.15 A milestone perspective presents a slightly different reality, since the nodes are tangible, which can point to re-arranged or even new tasks

3.6 Conclusion

In the challenging task of searching for link frameworks between the EIA and EMS processes, we analysed the processes individually and then together. A distinction between documents and tasks as the main sequence elements provided two alternative perspectives and corresponding representations of the processes, whether the processes were considered in a document-only, task-only, or an integrated view.

The action perspective of the integrated view provided the most complete and conservative framework for the EIA–EMS link. On the contrary, the milestone perspective of the integrated view provided a more tangible core sequence and hence an opportunity for a more flexible (re-)definition of tasks – a deviation from current standards which may be revolutionary for either process, with the advantages and disadvantages of any revolution.

Besides the process analysis presented in this chapter, there are other ways to seek for more link frameworks – for instance, by simile or by design. Thus, before making any commitments, there is much more exploration to be made – not so much conceptually, but perhaps more in practice.

4. Environmental Management Plans – Origins, Usage and Development

Bridget Durning

There is no formal requirement for an environmental management plan (EMP) within EIA legislation in the UK, but they are increasingly being used as a way of ensuring that mitigation measures proposed in the environmental statement (ES) are implemented on the ground (IEMA, 2008, 2011). Marshall (2004, p. [p. 141), who coined the phrase 'EMS-lite' to describe EMPs, also suggests that, by focusing on what is significant, they can act as the 'linking interface between the formal certified EMS and the EIA/project development'.

This chapter considers the origins of the concept of EMP. It briefly considers their effectiveness in delivering environmental protection through reference to a case study and their current usage. It concludes by reflecting on how EMPs can form the key link between EIA and EMS and how this can be further enhanced.

4.1 Origins

The concept of EMP has evolved over the last twenty years. The World Bank introduced environmental mitigation plans into its environmental assessment (EA) operating directive in 1991 (World Bank, 1993) as part of its focus on managing the environmental impacts of its investment. The concept evolved within World Bank documentation over the next few years to be referred to as 'environmental mitigation or management plan' and then into 'environmental management plan' (EMP) with guidance on what should be covered in the EMP being issued in January 1999. The broad aim of an EMP was to 'provide an essential link between the impacts predicted and mitigation measures specified within the EA report, and implementation and operational activities' (World Bank, 1999b, p. 1).

Hickie and Wade (1997) document the development of the concept of 'environmental action plans' (EAP) by the Environment Agency (the environmental regulatory agency in England and Wales) during the early 1990s as a way to strengthen their environmental assessment process. Prompted partly by the findings of Sadler (1996) but also in response to other specific needs of the Environment Agency, the aims of the EAP were described by Hickie and Wade (1997) as: to confirm details of environmental parameters and constraints for working in nature conservation areas; summarise environmental issues and constraints for the design team; clarify implementation and monitoring of environmental constraints and mitigation measures by agency staff; how any post-ES changes would be assessed and approved; how objectives and targets for successful post-project appraisal would be identified. They stated that EAPs detailed 'how the protection, conservation, mitigation and enhancement measures for the project will be delivered by the Environment Agency and its contractors' (Hickie and Wade, 1997, p. 789). A description of the current usage of EAPs by the Environment Agency is given in the example of EIA–EMS link in flood risk management in Chapter 10 (section 10.3.1).

Table 4.1 Comparison of requirements within early guidance on EMP

World Bank (1999b) – Aspects to be typically addressed within EMPs	**Hickie and Wade (1997) – Elements of an EAP in EA in England and Wales**
	A : Management and monitoring for final design and delivery of the project in accordance with the ES
Summary of impacts Description of mitigation measures	(1) summary of Environmental Assessment Process and the environmental constraints to be taken into account in terms of protection, conservation, mitigation and enhancement measures.
	(2) management of change in project design and implementation in relation to environmental impact;
Institutional arrangements	(3) communication programme to network in-house staff; engineering consultants and contractors; residents; landowners; public; user groups; and conservation bodies;

(Continued on next page)

World Bank (1999b) – Aspects to be typically addressed within EMPs	Hickie and Wade (1997) – Elements of an EAP in EA in England and Wales
Description of monitoring programme	(4) commitment to staff resourcing and procedures, normally a project Environmental Assessment Officer (as an independent member of the project team) and an Environmental Clerk of Works (as part of the supervising Resident Engineering team); Environmental Protection Schedules (EPS) to be checked by the Environmental Clerk of Works on a weekly basis; and Environmental Incident Forms and an associated reporting and follow-up system;
	(5) environmental assessment quality assurance system.
	B: Objectives and targets for each environmental constraint
	(1) objective;
	(2) implementation statement;
Implementation schedule and reporting procedures	(3) targets for objective (to be reviewed at post-project appraisal stage and remedial works instigated if necessary).
	C: Summary of environmental specifications required in engineering contract
	(1) contractors workmanship including procedures and limitations;
Cost estimates and sources of funds	(2) materials specifications.
	D: Drawing showing all constraints and comments

Key aspects of the two processes are shown in Table 4.1. Although titled slightly differently and evolving in response to differing operational needs there were some commonalities between the World Bank and Environment Agency processes. This is probably not too unexpected: Hickie and Wade acknowledged the evidence of similar approaches to their 'environmental project management' process (of which EAPs were a key part) being developed elsewhere, including the World Bank (previously described) and the Hong Kong Environmental Protection Department – see Sanvicens and Baldwin (1996), referenced in Hickie and Wade (1997). The literature of the time also documents other examples, including Bailey (1997) who referred to the long established importance of environmental management in Western Australia and the use of 'environmental management programmes' (also abbreviated to 'EMP') and Barnes and Lemon (1999) who documented an early example of an EMP for transport infrastructure development in Canada. The Barnes and Lemon example is notable by not only being one of the earliest documented examples of use of EMP, but also for it being a proponent driven process, which extended beyond the construction phase to encompass the full life span of the operational development.

The 'mood' at the time had been set by the international study into the effectiveness of EA (reported in Sadler (1996, p. 126)) which had emphasised that without going beyond the EIA phase 'the process [of environmental assessment] risks becoming a pro-forma exercise rather than a meaningful exercise in environmental management' although Holling, as far back as 1978, had observed that environmental management should be an ongoing process and 'not a one-time prediction of impacts' (Holling, 1978, p. 133). Table 4.2 shows the key aspects of the EMP from Barnes and Lemon (1999) and two other anonymised examples of practice from this time.

Table 4.2 Examples of early (1997–1999) EMPs of infrastructure projects

	Transport	**Energy**	**Water**
Country	Canada	UK	Eastern Europe
Author	Canadian proponent	UK proponent	UK consultant
Aim	Umbrella document which describes how the developer is managing all environmental aspects of the project throughout the construction phase and the subsequent 35-year period of private ownership	Operational manual for implementing appropriate environmental controls and monitoring procedures within the construction phase of development	Developed in response to analysis of impacts – series of tables, not standalone document

(Continued on next page)

	Transport	Energy	Water
Contents	Environmental protection plan; Environmental effects monitoring programme; Public information programme; Environmental reporting procedures; Contingency and emergency response plan	Environmental policy; Details of specific mitigation measures; Environmental monitoring; Implementation responsibilities; Community liaison; Legislation (regulation/consents/permits) register; Recording; Emergency measures	Issue/aspect; Mitigation/monitoring activity; Responsibility; Total phase 1 incremental cost; Timing; Training requirements; Institutional roles and responsibilities in relation to each of the mitigation and monitoring measures; Provisional cost estimates and timing in relation the main rehabilitation schemes; Any additional training needed

4.2 Case Study to Consider Effectiveness

This section considers a case study from the late 1990s in which an EMP process was adopted to manage the environmental impacts of the development during the construction phase. Specific details of this case study are provided in Broderick and Durning (2006); Broderick et al. (2010) and consequently only general aspects are presented here with the main focus being on the elements of the EMP and their effectiveness.

The development comprised a pipeline through open countryside in southern England. The topography along the route was gently rolling, although some areas of high ground lay on the fringes of the route and it also passed through a river valley system. The broad purpose of the (proponent derived) EMP was to:

- provide a mechanism for ensuring that measures to mitigate potentially adverse environmental impacts are implemented;
- ensure that standards of good construction practice are adopted throughout the construction of the pipeline;
- provide a framework for mitigating impacts that may be unforeseen or unidentified until construction is underway.

To be successful an EMP should evolve throughout the life of the project and this practice was adopted in the example with refinements occurring due to, for instance, design changes or comments from stakeholders. The EMP identified eleven activities which may have given rise to potential impacts during the construction of the pipeline and for which mitigation measures were required. Although the EMP detailed the mechanisms through which the impacts were to be addressed during construction of the pipeline and the responsibilities for meeting them, it was the contractor who was required to provide method statements of the details of the actions to be taken. The method statements had to demonstrate how compliance with the requirements of the EMP where to be achieved and specify the names of the individual people would be charged with achieving and monitoring compliance. The EMP also provided a framework for compliance auditing to ensure that its aims were being met. As it formed part of the commercial contract for the contractor during the construction of the pipeline, the proponents required that inspections and audits were undertaken to ensure that the plan was being implemented. In addition to any audits the contractor may have undertaken, the proponent also commissioned consultants to undertake periodic site audits. Where problems were identified corrective actions were required to be undertaken (these could have been further direct mitigation, changes to procedures or additional training).

As described in Broderick and Durning (2006), a visit to the site some nine years later showed very little evidence of a legacy of the pipeline. In one field the route could be identified by a darker green swath of grass, suggesting that restoration had slightly changed the drainage capacity in that area. The most obvious evidence was in the restored hedge where the extent of growth had not yet reached that equivalent to the original on either side of the pipeline route.

4.3 Current Usage on EMP in Practice

In order to consider the evolution of EMP practice, a scoping exercise comprising a comparative assessment of a randomised sample of ten EMPs produced between 2003 and 2010, was undertaken. The results are shown in Table 4.3, and they are are anonymised to focus on comparative practice and not on issues of scale, type or location of development.

Table 4.3 Comparison of a selection of EMPs produced between 2003 and 2010

EMP	Aim	Content
Date: 2003; Country: UK; Type of Document: Contractors (outline) EMP; Developed by: Consultant	Describes how construction activities will be planned and implemented in accordance with requirements of decision related documents (ES, Planning Inspector/Secretary of State decision letter); also describes how checking, monitoring and auditing will be carried out	Policy; Organisation – roles and responsibilities of environmental staff; Planning of activities: register of environmental impacts and commitments (mitigation options); environmental action plan; project risk register; register of environmental standards and legislation; objectives and targets Implementation of activities – training and awareness; communication; document control; operation control; supply chain requirements; project emergency plan; Checking and corrective action: monitoring; internal auditing; Document reviewed annually
Date: 2004; Country: Middle East; Type of Document: EMP – requirements of proponents EMS incorporated into the ES and EMP; Developed by: Consultant developed for proponent	Operational manual for implementing appropriate environmental controls and monitoring processes during the demolition, construction and operational phase of development	Policy and management responsibilities; Legislation register; Details of specific mitigation measures in each phase; Environmental monitoring; Method for recording environmental management during the demolition and construction phase; Emergency measures and procedures; Environmental auditing
Date: 2007; Country: Middle East; Type of Document: Topic specific management plans; Developed by: Consultant	Not stated	Nature of the work; Objectives; Responsibility for EMP; Choice of subcontractor; Schedule of work; Location; Responsibility of EMP monitoring and commitment

(Continued on next page)

EMP	Aim	Content
Date: 2009; Country: UK; Type of Document: Chapter with ES on environmental management which contains the EMP; Developed by: Consultant	Provides details of the measures to be adopted to minimise the potential for pollution	Procedures; Training; Communication; Monitoring; Environmental mitigation measures during construction, operation, decommissioning and restoration
Date: 2010; Country: Eastern Europe; Type of Document: Construction Environmental Control Plan (CECP); Developed by: Proponent	Describes the environmental management programme for project construction activities; objective s to define the requirements for compliance with regulations/permits/consents, the project EIA and EMP	Project and site description; Responsibilities; Environmental management controls; Environmental requirements and mitigation; Legislation register; Impacts and aspects register; Environmental management programme (objectives and targets)
Date: 2010; Country: Eastern Europe; Type of Document: Environmental and social management plan (ESMP); Developed by: Consultant	Describes general approach to how the contractors will conduct construction operation activities; describes the monitoring and evaluation processes	Purpose of ESMP; Environmental management; Organisation; Relevant legislation/permits/consents; Implementation (training, communication, document control, operational control); Environmental impacts and mitigation options; Monitoring; auditing; review; Specialist procedures and guidance notes; Emergency plan

(Continued on next page)

EMP	Aim	Content
Date: 2010; Country: Eastern Europe; Type of Document: Environmental and Social Management Plan (ESMP); Developed by: Consultant	Provide a mechanism for ensuring that measures to mitigate potentially adverse environmental impacts are implemented; that good standards during construction are adopted; provide a framework for mitigating impacts that may be unseen or unidentified until construction is underway; provide assurance to third parties that their requirement to respect to environmental performance will be met; to provide framework for compliance auditing and inspection	Processes to be carried out on site; ESMP; Environmental management; Objectives and actions required; Organisation (roles and responsibilities); Legal and other requirements; Environmental impacts and aspects; Summary of monitoring; Specific control measures, programme for management and monitoring for aspects; Assessment of environmental protection activities and reporting: Training; communication; Document control; operation control; responsibilities; auditing programme; review; specialist procedures and guidance notes Reporting; emergency plan
Date: 2010; Country: Eastern Europe; Type of Document: EMP; Developed by: Proponent	To highlight environmental risk posed by the company's activities: EMP describes the key components of the environmental management system that reveals in details the Company's environmental strategy	Introduction; Environmental management plan; Objectives and actions required; Legal and other requirements; Significant environmental impacts and aspects; Processes to be carried out at the site; Chapters on specific aspects; Significant environmental aspects and impacts; objectives and actions required, control measures; water management programme; monitoring; Assessment of environmental protection activities and reporting

(Continued on next page)

EMP	Aim	Content
Date: 2010; Country: Middle East; Type of Document: Chapter within an ES which contains Environmental management and monitoring plan (EMMP); Developed by: Consultant	Outlines the environmental goals of the project; aims to provide the direction for the control, mitigation and management review process; provides direction for the control, mitigation, monitoring, report and auditing necessary to prevent or mitigate potentially adverse environmental effects, incidence and emergencies that may occur during the construction and operational periods	Introduction; Construction phase environmental management plan; Operational environmental; Training; Performance improvement and learning; Environmental monitoring
Date: 2010; Country: North America; Type of Document: EMP; Developed by: Consultant	Primary objective is to ensure that adverse environmental effects resulting from the implementation of the project are minimised; also needed to meet development approval condition which required an EMP to be submitted for review and approval prior to commencement of work	Introduction; Environmental requirements; Project description and purpose; Environmental protection plan; Contingency and emergency response plans; Key contacts; Follow-up programme

This scoping study shows wide variation in the remit of the EMP (e.g. construction environmental management plans, environmental and social impact plans) with some being standalone documents and some chapters within environmental statements. Some are also specific in the monitoring and auditing measures to be adopted in the EMP whilst others are less specific. Some also do have elements of being an 'EMS-lite' (Marshall, 2004) by containing some of the key elements of an EMS, such as policy, procedures, auditing and review. Others have the requirement to link to the development proponents EMS. Practice is therefore relatively diverse. However, it seems that EMPs are popular with practitioners: a report on a survey by the UK based Institute for Environmental

Management and Assessment (IEMA) on EIA effectiveness demonstrated the popularity of the use of EMPs in the UK: it reported that 80 per cent of its survey respondents stated they would welcome the inclusion of EMP as a mandatory requirement within EIA legislation (IEMA, 2011). The lack of legislative requirement has led to guidance being produced by a range of organisations – for instance, World Bank (1999b), Provincial Government of the Western Cape (2005), IEMA (2008), The Highland Council (2010). All the guidance emphasises that EMPs should be targeted to the nature of the development (which to some extent explains the variation in practice) and should be regularly reviewed. It is certainly likely that if not regularly reviewed then EMP risks becoming another 'tick box' approach to comply with the consent process and not to manage the environmental impacts of the development.

4.4 Reflecting on How EMPs Can Form the Key Link between EIA and EMS

The use of EMP has the appearance of becoming an established process: the regular presentation of examples of practice at the International Association for Impact Assessment (IAIA) annual conferences suggests their use across a range of sectors.

Marshall (2004) suggested that EMP could be considered as 'EMS-lite' and certainly they currently offer a way of managing the environmental impacts of development at a number of stages. One of the key elements of EMS is continual improvement and this would also need to be the case in 'EMS-lite'. This element of constant review and therefore adaptation, is already built into the World Bank guidance on EMP. By being considered part of a continuum, EMP, as part of a systematic environmental project management process which is regularly monitored and reviewed, allows for improvement in the identification and control of impacts in relation to the design aspects of a development, as envisaged by Hickie and Wade in 1997. The recent review of EIA practice in the UK (IEMA, 2011) emphasised this through its codification of the key stages of a successful EMP:

1. Involvement of the proponent, construction teams and contractors during the formulation of mitigation when planning the development to ensure:

- The mitigation is deliverable;
- Costs of mitigation actions are factored into the detailed design stage, whilst the construction budget is still being developed; and
- There is buy-in and commitment when developing mitigation at the EIA stage, to increase likelihood of effective implementation.

2. Involvement of competent environmental professionals in the design and specification of mitigation actions to ensure:
   - That the requirements are very clear; and
   - To improve the chances of successful delivery;
3. Ensuring mitigation measures identified whilst planning the development have been formulated with the involvement of relevant stakeholders, or a clear timetable and process for further consultation post-consent is set out (where mitigation requires more detailed design post-consent);
4. Clearly identified and sufficiently detailed mitigation measures in the pre-consent phase (a framework or draft EMP prepared alongside an ES is helpful as it can set out additional detail in preparation for a full EMP);
5. Ensuring mitigation proposals are identifiable in documents accompanying the application for development consent;
6. Ensuring mitigation measures are presented as elements that the proponent would be willing to have included in the final consenting documentation (IEMA, 2011, p. 84).

Figure 4.1 provides a useful summary of the environmental project management process and how EMP fit within the process.

One of the challenges in the use of EMP is maintaining the adaptive element, particularly in relation to large and complex projects. As Faith-Ell and Arts (2009, p. 5) observed in their reflection on the role of private sector investment in large scale infrastructure projects: 'as early market involvement will usually lead to longer-term contracts, these contracts will have to deal [with] changing contexts – new spatial and environmental developments, new techniques, new regulations and policies etc. As a consequence, there is need for some form of environmental management responsiveness.' They suggest adopting processes which are in essence those advocated by IEMA as key – that is, establishment of a system to manage environmental impacts, with careful monitoring of contract requirements and periodic evaluation of environmental performance measures.

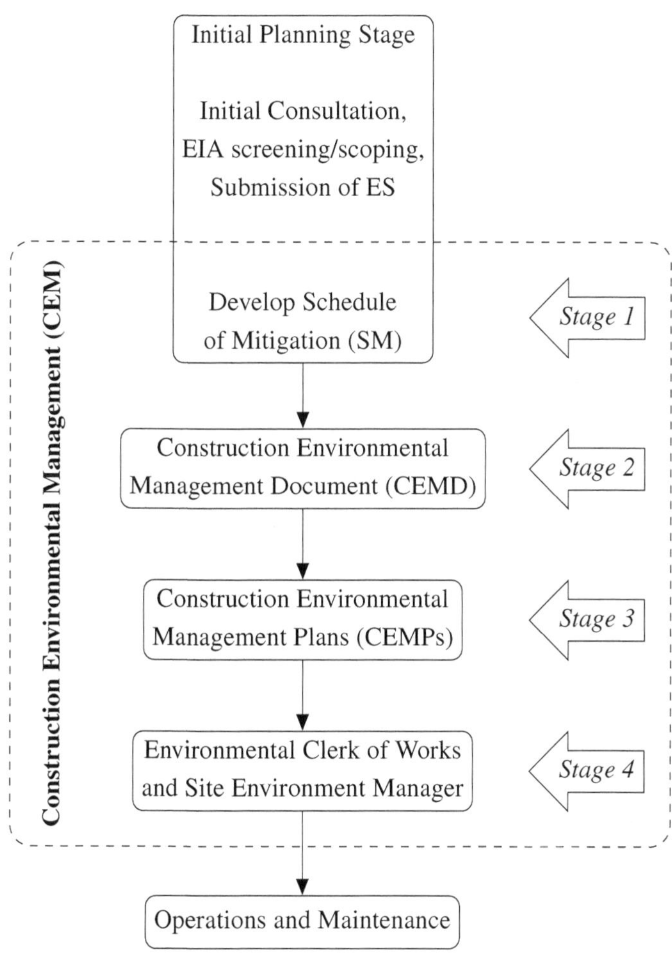

Figure 4.1 Project Environmental Management Process – source: The Highland Council (2010)

Based on evidence reviewed, apparently there has been a tendency to concentrate the use of EMP at the construction stage of the development. This was apparent at the early stages of development of the concept when, due to the nature of the projects being delivered by the Environment Agency (in England and Wales), the EAP concluded at the point when the development becoming operational. What is possibly still an issue which needs further consideration are those developments where the EMP needs to link into the long term operation and ultimate decommissioning stage of a development. The guidance for the World Bank EMP process states that the EMP should continue through to the operational stage of developments and be regularly reviewed. Figure 4.2 shows an idealised linkage through EIA to EMP to EMS.

4.5 Thoughts on Further Enhancement

The use of EMPs has increased over the last few years in the UK – for instance, IEMA (2008, 2011) – although the focus has mainly concentrated on management and mitigation of the construction effect of development. There is still much room for improvement in environmental management planning. Whilst many suggest change in legislation may achieve this, research carried out by Mordue (2008) using evidence from numerous countries with different legislative systems and cultures, showed that even when post EIA-monitoring is a statutory requirement there is considerable room for improvement: regulation alone is not sufficient to instigate an effective monitoring regimes. Mordue (2008) identified many reasons for this, including lack of resources, lack of expertise and inadequate techniques.

Marshall et al. (2005) have suggested that practitioners should be the ones to take forward improvement in the practice of impact monitoring and management. However, this is likely to depend on the forum in which practice is occurring – for instance, by proponent or by commercial consultancy, as the latter faces commercial pressures to undertake only those tasks for which they are paid. Guidance by professional bodies (e.g. IEMA (2008)) can give rise to the development of communities of practice (Lave and Wenger, 1991) and these have potentially been instrumental in the rise in use of EMP in the UK. But possibly more is needed. There are arenas where practice is being driven forward through other means and this is most strongly observed through the

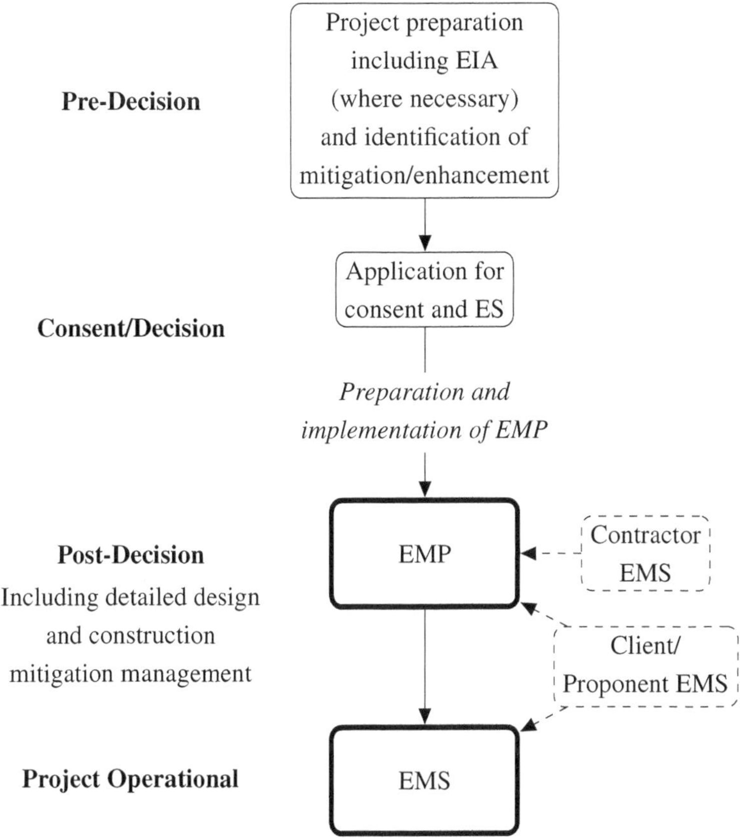

Figure 4.2 Idealised EMP to EIS linkage – source: IEMA (2011)

influence of the Equator Principles. Launched in 2003, these are a voluntary set of guidelines for financial institutions which form a 'credit risk management framework for determining, assessing and managing environmental and social risk in project finance transactions' (Equator Principles, 2011). Based on the World Bank EHS Guidelines and IFC Performance Standards they have been adopted by over 70 financial institutions in 27 countries (Equator Principles, 2011). Research has shown that where used they do improve practice (Ferguson, 2008), although most strongly in areas relating to social/community aspects.

The establishment of the Equator Principles represents the manifestation of something approaching an epistemic community within one aspect of environmental impact assessment and management. Defined by Haas (1992, p. 3), epistemic communities are:

> a network of professionals with recognised expertise and competence in a particular domain and an authoritative claim to policy-relevant knowledge within that domain or issue-area'

Under Haas' definition, members of epistemic communities, although similar to those who form part of a disciplinary or professional group, have 'shared normative commitments'. This shared normative commitment allows them to develop knowledge and expertise which others practising in the field could refer to for guidance on appropriate methods to adopt in particular areas of practise. In light of the evidence that legislation is unlikely to change practice in the management of impacts of development post EIA, the possibility of the development of more epistemic communities is an area that would be worthy of further investigation.

5. ESIA Effectiveness Through Links to EMS

Martin Broderick

Environmental and social impact assessment (ESIA) is well established as an environmental policy instrument. As with any well established tool, there is ongoing debate about how effective it really is (Sadler, 1996; Carroll and Turpin, 2002; Turpin, 2010). Such studies have tended to focus on macro scale procedural aspects within national jurisdictions (Wood, 2003) with less attention being paid to the micro-scale more substantive outcomes, i.e. improving the projects operational and post-closure environmental performance. Therefore, a key question at the micro-scale project level is: will the ESIA, on its own, lead to the project (and impacted environment) being managed in an acceptable way?

Appropriately employed, ESIA is a key integrative element in environmental protection, but is only one element of that policy toolbox (Wood, 2003). Other elements include the monitoring and evaluation of the impacts of a project (which has been subject to ESIA) and the subsequent management of the environmental performance of that project. This process has been termed ESIA 'Follow-Up' (Morrison-Saunders and Arts, 2004b). The follow-up processes have the same goal as the key ESIA elements, namely to minimise the negative consequences of development and maximise the positive, with the emphasis on the action being taken to achieve this goal. It has been suggested by some authors that ESIA has little value unless follow-up is carried out: without it the process remains incomplete and the consequences of ESIA planning and decision making are unknown. By minimising the negative and maximising the positive outcomes, it is postulated that ESIA follow-up can provide the safeguard for environmental protection (Marshall et al., 2005). Through an exploration of the many drivers for ESIA development and differing methodological approaches, this chapter proposes that the most effective ESIA systems are those that use follow-up processes to tangibly link ESIA and EMS.

5.1 Drivers of ESIA Development

Increasing concerns in the developed economies in the 1960s about the impact of anthropogenic activities on the biophysical environment and on human health led to the development of the concept of ESIA and its adoption as a legally based decision support instrument to assess the environmental implications of proposed development. The diffusion of ESIA practice through legal instruments has been the primary driver of ESIA take-up (Wood, 2003), although it has also been given impetus by the development of voluntary codes and principles in industrial and financial sectors. Additionally, ESIA practitioners, through international and nationally dedicated impact assessment professional bodies, have produced a range of best practice guidance. Each of these different drivers is explored briefly in the following sections.

5.1.1 Legislative Drivers

ESIA is a systematic and integrated process first developed in the USA as a result of The National Environmental Policy Act (NEPA) which was signed into law on January 1, 1970 (Glasson et al., 2005). Acknowledging the decades of environmental neglect that had significantly degraded the nation's landscape and damaged the human environment, the law was established to: 'foster and promote the general welfare, to create and maintain conditions under which man and nature can exist in productive harmony, and fulfil the social, economic, and other requirements of present and future generations' (NEPA, 1969, Sec.101(a)). A number of nations rapidly followed suit (Wood, 2003): 1972 – New South Wales State, Australia, adopted 'an extremely rudimentary environmental impact policy'; 1973 – Canada approved a Federal Directive on ESIA; 1974 – Australia adopted an ESIA Act and New Zealand instituted ESIA procedures; 1976 – France and West Germany established an ESIA system; 1979 – the Netherlands adopted an ESIA system.

ESIA was internationally recognised at the UN Conference on Environment and Development, held in Rio de Janeiro in 1992. Principle 17 of the Final Rio Declaration stated: 'Environmental impact assessment, as a national instrument, shall be undertaken for proposed activities that are likely to have a significant adverse impact on the environment and are subject to a decision of a competent national authority.'

The European Communities ESIA Directive (85/337/EEC) came into force in 1985. It applies to a wide range of defined public and private projects (defined in Annexes I and II of the Directive) and has been amended three times (1997, 2003 and 2009) (Wood, 2003):

- Directive 97/11/EC brought the Directive in line with the UN ECE Espoo Convention on ESIA in a Transboundary Context. The Directive of 1997 widened the scope of the ESIA Directive by increasing the types of projects covered, and the number of projects requiring mandatory environmental impact assessment (Annex I). It also provided for new screening arrangements, including new screening criteria (at Annex III) for Annex II projects, and established minimum information requirements.
- Directive 2003/35/EC seeked to align the provisions on public participation with the Aarhus Convention on public participation in decision making and access to justice in environmental matters.
- Directive 2009/31/EC amended the Annexes I and II of the Directive by adding projects related to the transport, capture and storage of carbon dioxide (CO_2).

Formal EIA legislation was enacted in the UK in 1988, although as Glasson et al. (2005) shows a vast range of EIA legislation has been enacted in the UK since. Most recently the Planning Act 2008, which deals with Nationally Significant Infrastructure Projects (NSIPs), has aimed to put impact assessments at the heart of the consenting process. In 2010 a review of the EC Directive was launched with the objective of gaining an overall view on the functioning and effectiveness of the ESIA Directive and the need to amend the ESIA Directive appropriately. At the time of writing the outcome of the review is still awaited. The continuing absence of formal follow-up requirements in the EC Directive militates greatly against the effectiveness of ESIA in the EU. There are a number of countries that are acknowledged as having sophisticated systems of environmental controls, particularly in relation to EIA. For example:

The Netherlands through its Environment Management Act (EMA) (1994) and the EIA Decree VROM (1996) the Netherlands has been, for many years, regarded by many as the most effective in Europe (Glasson et al., 2005; Sadler, 1996; Wood, 2003). It has requirement for follow-up, although it is acknowledged that the process was slow to commence, with only a 'handful' being carried out in the first few years (Arts, 1998). The legislation has recently undergone 'modernisation' (Runhaar, 2011) with the removal of some of the mandatory review requirements which unfortunately brings elements of uncertainty on the future effectiveness of the legislation.

Hong Kong has clear formal regulations combined with carefully considered process and content details. The Environmental Monitoring and Auditing (EMA) component of Hong Kong's EIA Ordinance includes a requirement for an independent environmental checker to verify and certify systematically that mitigation measures proposed in the ES are fully and properly implemented. This has produced an effective and innovative ESIA follow-up system (EPD, 2002).

Canada EIA was formally introduced in Canada in 1973 by the federal Environmental Assessment and Review Process (EARP). In 1992, the Canadian Environmental Assessment Act was introduced to replace EARP and to strengthen EIA (Noble, 2010). EIA is also required under other laws: municipalities and corporations are subject to the EIA requirements of their respective provincial, territorial or land claim jurisdictions but are also subject to the Canadian Environmental Assessment Act if the federal government holds some decision making authority concerning the proposed development or the acceptability of its impacts. Noble also notes that informally, EIA is increasingly becoming a routine part of the environmental management and auditing systems of municipalities and corporations: early involvement of the public, and the public's sustained involvement throughout the process is a key part of Canadian law together with the monitoring and managing actual outcomes.

However, even in national regimes where EIA procedures are highly regarded (as described above) effectiveness is unfortunately often measured by procedural compliance rather than project outcomes. With ESIA practice now well into its second generation it is vital that legislative regime become 'mature' by embedding follow-up and monitoring in the legislative regimes.

5.1.2 Voluntary Codes/Principles

The World Bank

Whilst most multi-lateral development banks now have systems and guidance on ESIA, the World Bank first developed environmental guidelines back in 1988. These were superseded in 1999 by its Pollution Prevention and Abatement Handbook (World Bank, 1999a) and its first operational directive on ESIA (World Bank, 2011).

The Pollution Prevention and Abatement Handbook is specifically designed to be used in the context of the World Bank Group's environmental policies, (as set out in Operational Policy (OP) 4.01, 'Environmental Assessment', and related documents). It comprises three sections covering key policy lessons in pollution management, good-practice notes on implementation of policy objectives and detailed guidelines to be applied in the preparation of World Bank Group projects. The guidelines, which cover almost 40 industrial sectors, represent state-of-the-art thinking on how to reduce pollution emissions from the production process. In many cases, the guidelines provide numerical targets for reducing pollution, as well as maximum emissions levels that are normally achievable through a combination of cleaner production and end-of-pipe treatment.

The aim of its ESIA operational directive, which has been updated a number of times since 1999 (World Bank, 2011) is to ensure that projects proposed for funding by the bank are environmentally sound and sustainable and thus to improve its decision making. The Bank favours preventive measures over mitigatory or compensatory measures, whenever feasible and monitoring and follow-up are integral to the procedures.

International Finance Corporation (IFC)

IFC has a long historical record of having environmental and social standards. It adopted its Environmental and Social Safeguard Policies and its Disclosure

Policy in 1998 which was superceded in 2006, when IFC had applied the Policy and Performance Standards (PSs) on Social and Environmental Sustainability (IFC, 2006) to all investment projects to minimise their impact on the environment and on affected communities. The PSs are also used by other financial institutions on a voluntary basis. There are eight PSs):

1. Social and Environmental Assessment and Management System
2. Labour and Working Conditions
3. Pollution Prevention and Abatement
4. Community Health, Safety and Security
5. Land Acquisition and Involuntary Resettlement
6. Biodiversity Conservation and Sustainable Natural Resource Management
7. Indigenous Peoples
8. Cultural Heritage

IFC (2006) state:

> Performance Standard 1 establishes the importance of: (i) integrated assessment to identify the social and environmental impacts, risks, and opportunities of projects; (ii) effective community engagement through disclosure of project-related information and consultation with local communities on matters that directly affect them ; (iii) the client's management of social and environmental performance throughout the life of the project.
>
> Performance Standards 2 to 8 establish requirements to avoid, reduce, mitigate or compensate for impacts on people and the environment and to improve conditions where appropriate. While all relevant social and environmental risks and potential impacts should be considered as part of the assessment, Performance Standards 2 to 8 describe potential social and environmental impacts that require particular attention in emerging markets. Where social or environmental impacts are anticipated, the client is required to manage them through its Social and Environmental Management System consistent with Performance Standard 1.

Monitoring and follow-up are integral to PSs and are intimately linked to Environmental, Health and Safety (EHS) Guidelines (Ranking et al., 2008). The EHS Guidelines (IFC 2011) are described by the IFC as being: 'technical reference documents with general and industry specific examples of Good International Industry Practice (GIIP) . . . Reference to the EHS Guidelines by IFC clients is required under Performance Standard 3 whilst IFC uses the EHS Guidelines as a technical source of information during project appraisal activities'. The EHS Guidelines contain the performance levels and measures that are normally acceptable to IFC and are generally considered to be achievable in new facilities at reasonable costs by existing technology. For IFC-financed projects, application of the EHS Guidelines to existing facilities may involve the establishment of site-specific targets with an appropriate timetable for achieving them. The environmental assessment process may recommend alternative (higher or lower) levels or measures, which, if acceptable to IFC, become project- or site-specific requirements.

In my own experience on IFC-funded projects the monitoring and follow-up requirements of the PSs lead in a tangible way to better outcomes for all stakeholders involved.

The European Bank for Reconstruction and Development (EBRD)

The European Bank for Reconstruction and Development (EBRD) adopted its first environmental policy in 1991. This policy and the related 10 Performance Requirements (PRs) were updated in 2008 (EBRD, 2008). Bank-financed projects are expected to meet good international practice related to sustainable development. To help clients and/or their projects achieve this, the Bank has defined specific PRs for key areas of environmental and social issues and impacts as listed below:

1. Environmental and Social Appraisal and Management
2. Labour and Working Conditions
3. Pollution Prevention and Abatement
4. Community Health, Safety and Security
5. Land Acquisition, Involuntary Resettlement and Economic Displacement
6. Biodiversity Conservation and Sustainable Natural Resource Management

7. Indigenous Peoples
8. Cultural Heritage
9. Financial Intermediaries
10. Information Disclosure and Stakeholder Engagement

EBRD (2008) require all companies/institutions in receipt of funding to have a systematic approach to managing the environmental and social issues and impacts associated with the funded project so that it can comply with the Bank's Environmental and Social Policy. They describe such an approach as: 'A successful and efficient environmental and social management system is a dynamic, continuous process, initiated and supported by management, and involves meaningful communication between the client, its workers, and the local communities affected by the project or the client company. It requires a methodical systems approach comprising planning, implementing, reviewing and reacting to outcomes in a structured way with the aim of achieving a continuous improvement in performance.' Monitoring and follow-up protocols are a key part of the EBRD's PRs.

Equator Principles

The Equator Principles (2011) (EPs) are a financial industry benchmark for determining, assessing and managing social and environmental risk in project financing. The EPs are based on World Bank and IFC guidelines (described previously) and oblige committing financial institutions to finance projects only if it can be guaranteed that the social and ecological impact of projects are assessed (Broderick et al., 2010).

The EP Financial Institutions (EPFIs), of which there were 69 as of 16 March 2011, have consequently adopted these Principles in order to ensure that the projects they finance are developed in a manner that is socially responsible and reflect sound environmental management practices. By doing so, negative impacts on project-affected ecosystems and communities should be avoided where possible and if these impacts are unavoidable, they should be reduced, mitigated and/or compensated for appropriately. The EPFIs review these Principles from time-to-time based on implementation experience in order to reflect and learn from emerging good practice.

There are 10 Principles, but Principle 4 (Action Plan and Management

System) is the key relevant one for linking ESIA and EMS and is reproduced below (emphasis added):

> For all Category A and Category B projects located in non-OECD countries, and those located in OECD countries not designated as High-Income, as defined by the World Bank Development Indicators Database, the borrower has prepared an Action Plan (AP) which addresses the relevant findings, and draws on the conclusions of the Assessment. *The AP will describe and prioritise the actions needed to implement mitigation measures, corrective actions and monitoring measures necessary to manage the impacts and risks identified in the Assessment.* Borrowers will build on, maintain or establish a Social and Environmental Management System that addresses the management of these impacts, risks, and corrective actions required to comply with applicable host country social and environmental laws and regulations, and requirements of the applicable Performance Standards and EHS Guidelines, as defined in the AP.
>
> For projects located in High-Income OECD countries, EPFIs may require development of an Action Plan based on relevant permitting and regulatory requirements, and as defined by host-country law. The Action Plan may range from a brief description of routine mitigation measures to a series of documents (e.g., resettlement action plan, indigenous peoples plan, emergency preparedness and response plan, decommissioning plan, etc.). The level of detail and complexity of the Action Plan and the priority of the identified measures and actions will be commensurate with the project's potential impacts and risks. Consistent with Performance Standard 1, the internal Social and Environmental Management System will incorporate the following elements: (i) Social and Environmental Assessment; (ii) management programme; (iii) organisational capacity; (iv) training; (v) community engagement; (vi) monitoring; and (vii) reporting.

In my own experience the application of the EPs in the mining sector of low income economies points the way forward to improving the effectiveness of Socio Economic Impact Assessment in high income economies, as defined by the Organisation for Economic Co-Operation and Development (OECD). The Equator Principles and the International Finance Corp (IFC) Performance Standards, on which the principles are based, have increased the attention given to social and economic issues within Environmental & Social Impact Assessment (ESIA) in the low income economies of the world (Broderick, 2011). Coupled with the requirement for monitoring and follow-up in the EPs, projects supported by EPFIs are clearly resulting in more effective outcomes.

5.1.3 Good Practice Guidance

International Council for Mining and Metals (ICMM)

ESIA is typically associated with the exploration and feasibility stages of the mining project cycle, whereas EMSs are more closely associated with operations and mine closure. The ICMM has produced good practice guidance for its members which emphasises the importance of the systems, tools and processes of ESIA and EMS being applicable at any of the mining development stages i.e. development, operation and closure (ICMM, 2009). For example, they highlight the relevance of implementing an EMS during exploration to provide a framework for identifying and managing impacts at this early stage. Similarly, the determination of the significant aspects for an EMS may require the application of the evaluation and assessment stages of ESIA. ICMM views ESIA as a process for managing environmental and social impacts rather than an exercise solely linked to permitting requirements.

The mining sector is leading the development of good practice by linking ESIAs and EMS and coupled with the EPs the mining sector is clearly leading to more effective outcomes – see Figure 5.1.

Design Manual for Roads and Bridges (DMRB)

The DMRB was introduced in 1992 in England and Wales and subsequently in Scotland and Northern Ireland (Department for Transport Highways Agency, 2011). It provides a comprehensive system which accommodates all current standards, advice notes and other published documents relating to the design, assessment and operation of trunk roads (including motorways). Environmental management/Action plans are a key component for the operational aspects of roads and highways environmental performance (Department for Transport Highways Agency, 2011, Vol.11) – see Figure 5.2.

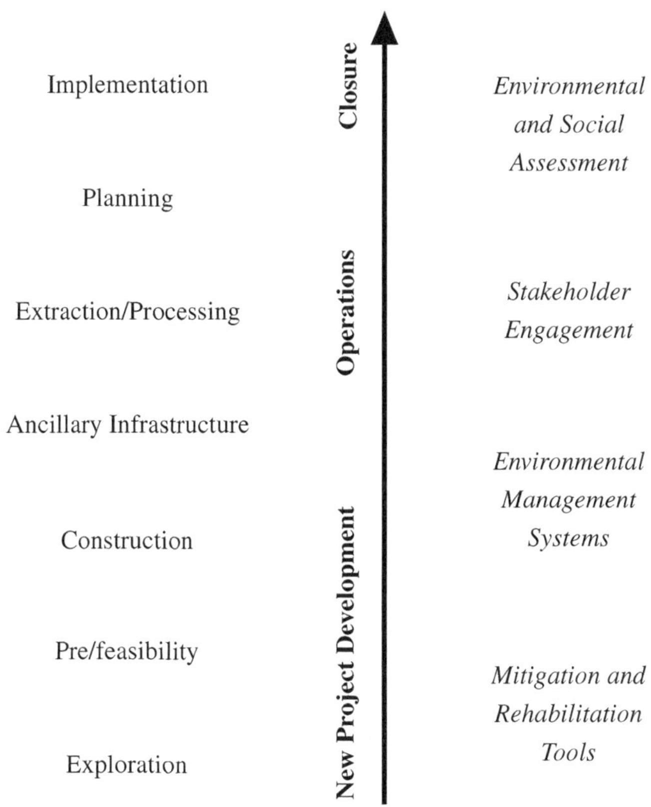

Figure 5.1 ESIA and EMS integration in mining sector; systems, tools and processes (right column, in italic) can apply at any stage of the project – source: ICMM (2009)

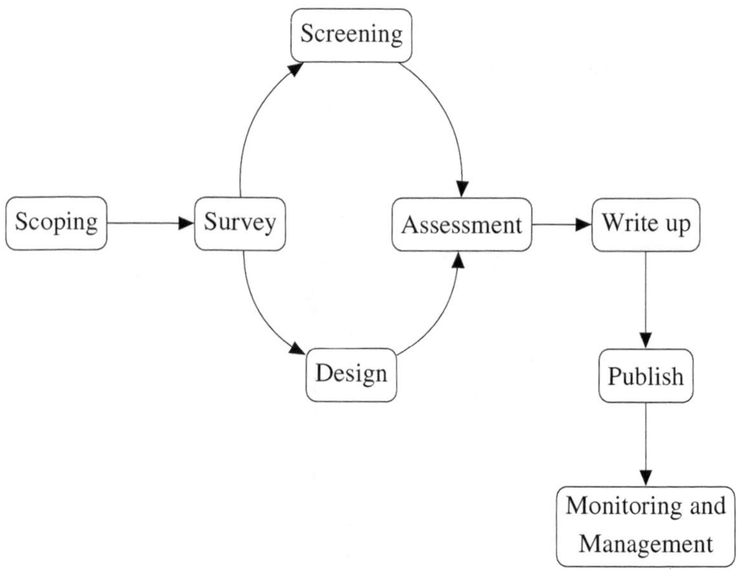

Figure 5.2 The EIA process – source: Department for Transport Highways Agency (2011)

5.1.4 Practitioners

The International Association for Impact Assessment (IAIA)

IAIA is a professional body which describes itself as the 'leading global network on best practice in the use of impact assessment'. IAIA was established in 1980 to bring together all involved in impact assessment, i.e. researchers, practitioners and users of various types of impact assessment from all parts of the world; it therefore involves many disciplines and professions, e.g. 'corporate planners and managers, public interest advocates, government planners and administrators, private consultants and policy analysts, university and college teachers and their students'. IAIA's goals are:

1. to develop approaches and practices for comprehensive and integrated impact assessment,
2. to improve assessment procedures and methods for practical application,
3. to promote training of impact assessment and public understanding of the

field,

4. to provide professional quality assurance by peer review and other means, and
5. to share information networks, timely publications, and professional meetings.

The Institute for Environmental Management and Assessment (IEMA)

IEMA in the UK, the Institute for Environmental Management and Assessment (IEMA) have also included follow-up as part of their best practice guidelines. IEMA state that follow-up is one of the most important parts of the EIA process as it helps determine whether EIA makes a difference in terms of improved environmental protection (IEMA, 2004).

However, despite these recommendations follow-up remains the one stage of the EIA process that is under-utilised worldwide. Below are the 'Best Practice' principles in EIA which were issued by the IAIA and IEMA (IAIA, 1999):

Purposive the process should inform decision making and result in appropriate levels of environmental protection and community well-being.

Rigorous the process should apply 'best practicable' science, employing methodologies and techniques appropriate to address the problems being investigated.

Practical the process should result in information and outputs which assist with problem solving and are acceptable to and able to be implemented by proponents.

Cost-effective the process should achieve the objectives of ESIA within the limits of available information, time, resources and methodology.

Efficient the process should impose the minimum cost burdens in terms of time and finance on proponents and participants consistent with meeting accepted requirements and objectives of ESIA.

Focused the process should concentrate on significant environmental effects and key issues; i.e., the matters that need to be taken into account in making decisions.

Adaptive the process should be adjusted to the realities, issues and circumstances of the proposals under review without compromising the integrity of the process, and be iterative, incorporating lessons learned throughout the proposal's life cycle.

Participative the process should provide appropriate opportunities to inform and involve the interested and affected publics, and their inputs and concerns should be addressed explicitly in the documentation and decision making.

Interdisciplinary the process should ensure that the appropriate techniques and experts in the relevant biophysical and socioeconomic disciplines are employed, including use of traditional knowledge as relevant.

Credible the process should be carried out with professionalism, rigor, fairness, objectivity, impartiality and balance, and be subject to independent checks and verification.

Integrated the process should address the interrelationships of social, economic and biophysical aspects.

Transparent the process should have clear, easily understood requirements for ESIA content; ensure public access to information; identify the factors that are to be taken into account in decision making; and acknowledge limitations and difficulties.

Systematic the process should result in full consideration of all relevant information on the affected environment, of proposed alternatives and their impacts, and of the measures necessary to monitor and investigate residual effects.

Inherent in many of these best practice principles is an aim to ensure that the EIA process is an effective environmental management tool. In particular, research suggests that the basic principles, which include phrases such as 'rigorous', 'interdisciplinary' and 'systematic' can all be associated with achieving good environmental management (Harmer, 2005). Within the guidelines set out by the IAIA the need for follow-up has been specifically recognised as an essential operating principle, in order:

- To ensure that the terms and conditions of approval are met.
- To monitor the impacts of the development.
- To monitor the effectiveness of mitigation measures.
- To strengthen future EIA applications and mitigation measures.
- To undertake environmental audit and process evaluation to optimise environmental management (IAIA, 1999).

5.1.5 Methodologies

Although legislative requirements and practices vary around the world (Wood, 2003; Glasson et al., 2005), the fundamental components of an ESIA are listed in below (emphasis added):

Screening to determine whether or not a proposal should be subject to ESIA and, if so, at what level of detail.

Scoping to identify the issues and impacts that are likely to be important and to establish terms of reference for ESIA.

Baseline Studies to characterise the area prior to development.

Examination of alternatives to establish the preferred or most environmentally sound option for achieving the objectives of a proposal.

Impact analysis to identify and predict the likely environmental, social and other related effects of the proposal.

Mitigation and impact management to establish the measures that are necessary to avoid, minimise or offset predicted adverse impacts and, where appropriate, to incorporate these into an environmental management plan or system.

Evaluation of significance to determine the importance or acceptability of residual impacts that cannot be mitigated.

Preparation of EIS or report to document the impacts of the proposal, the significance of effects, and the concerns of the interested public and the communities affected by the proposal.

Review of the EIS to determine whether the report meets its terms of reference, provides a satisfactory assessment of the proposal(s) and contains the information required for decision making.

decision making to approve or reject the proposal and to establish the terms and conditions for its implementation.

follow-up *to ensure compliance with the terms and conditions of approval; to monitor the impacts of development and the effectiveness of mitigation measures; and, where required, to undertake environmental audit and process evaluation to strengthen future ESIA applications and mitigation measures and to optimise environmental management.*

5.2 ESIA Follow-up: The Key to Linking ESIA and EMS

EIA follow-up can be simply defined as the: 'monitoring and evaluation of the impacts of a project or plan (that has been subject to EIA) for management of, and communication about, the environmental performance of that project or plan' (Morrison-Saunders and Arts, 2004b). Thus, EIA follow-up comprises four elements (Arts et al., 2001):

Monitoring the collection of activity and environmental data both before (baseline monitoring) and after activity implementation (compliance and impact monitoring).

Evaluation the appraisal of the conformance with standards, predictions or expectations as well as the environmental performance of the activity.

Management making decisions and taking appropriate action in response to issues arising from monitoring and evaluation activities.

Communication informing the stakeholders about the results of EIA follow-up in order to provide feedback on project/plan implementation as well as feedback on EIA processes.

Follow-up is essential for determining the outcomes of EIA. By incorporating feedback into the EIA process, follow-up enables learning from experience to occur. It can and should occur in any EIA system to prevent EIA being just a pro forma exercise (Morrison-Saunders et al., 2007).

5.3 Summary

This chapter has described the many drivers for ESIA development and summarised the different methodological approaches.

A key question at the microscale project level is: will the ESIA on its own lead to the project and impacted environment been managed in an acceptable way? The characteristics of an advanced effective ESIA system is that it is linked to an EMS through ESIA follow-up procedures, e.g. Hong Kong, Netherlands, ICMM, DMRB, WB/IFC and EPs. Appropriately employed, ESIA is a key integrative element in environmental protection but is only one element of that policy (Wood, 2003). Follow-up has the same goal as ESIA, namely to minimise the negative consequences of development and maximise the positive. The emphasis is on action taken to achieve this goal. ESIA has little value unless follow-up is carried out because without it the process remains incomplete and the consequences of ESIA planning and decision making will be unknown. By minimising the negative and maximising the positive outcomes, ESIA follow-up can ensure that the outcomes of the ESIA process are successfully incorporated to operational EMSs ensuring substantive and sustainable outcomes for projects (Marshall et al., 2005). It is therefore proposed that the most advanced and effective ESIA systems are those that tangibly link ESIA and EMS through the mechanism of ESIA follow-up.

6. EIA–EMS Link in Vietnam

Dao Mai Anh and Clive Briffett

Through a Vietnam case study we explore the potential role of utilising the ISO 14001 Environmental Management System (EMS) in association with Environmental Impact Assessment (EIA) requirements for development of a framework of firm compliance behaviour with regard to environmental laws and regulations. It first presents the background to the study and a country overview followed by outlining EIA and EMS practice in Vietnam. It then discusses the determinants of firm compliance behaviour with environmental laws and regulations and finally concludes with discussion of the role of EMS in implementing EIA follow-up requirements. The chapter draws from recent studies (Dao et al., 2009; Dao and Ofori, 2010).

6.1 Background to Study

For the last two decades, Vietnam has shown steady economic growth but has paid the price for this development with the continuing degradation of the environment. Air and water pollution and solid waste treatment has become a social issue, particularly in urban areas, where, as a result of economic development, people and factories are concentrated. The government has coped with this situation by establishing environmental laws and regulations starting with the enactment of the Law on Environmental Protection (LEP) in 1994. However, both personnel and budgetary foundations of environmental administrative organisations (central or local), are vulnerable and are not adequate for enforcing the environmental laws and regulations effectively.

The country has introduced specific environmental regulations designed to deal with water pollution, air pollution, and industrial waste as these are the country's principal environmental challenges. Companies intending at least to implement environmental conservation measures necessary to satisfy legal requirements have found many difficult problems with promoting environmental conservation measures in Vietnam. The responses to environmental problems have been diverse within the business community ranging from excellent performance, to those with no control measures at all. Unfortunately, those that have a bad environmental performance comprise the majority of businesses.

In the industrial sector, the existence of old production facilities and state-owned enterprises with inadequate financial strength for implementing pollution control measures is a problem that cannot be ignored. With an increase in the number of private businesses, including foreign companies, the composition ratio of state-owned enterprises is now down to about 40 per cent in the mining and manufacturing sectors (General Statistics Office of Vietnam, 2006), but these enterprises are implementing almost no pollution control measures. About 75 per cent of industrial estates located across the country are equipped with no central wastewater treatment facilities or other environmental protection equipment, except for those with strong financial resources like Vietnam-Singapore Industrial Park in Binh Duong and big multi-national companies (MNCs), especially Japanese companies (Le and Nguyen, 2001). Other than a number of the foreign companies that are active in environmental protection, many businesses are implementing almost no emission and wastewater control measures. When it comes to industrial waste, especially hazardous industrial waste, which is expected to become a serious environmental issue in Vietnam, there are now no facilities within the country that can treat and dispose of it as required by law. Solving such an issue will become a tough challenge for the country.

Almost all foreign companies have executed positive environmental conservation measures based on cost-benefit calculations which include fear of fines and penalties. Those with high environmental awareness implement conservation measures as part of their normal corporate activity. This is partly because many of their parent companies are established firms that promote similar environmental conservation measures based on their global environmental policies, but also largely because their executives, mostly foreigners, have experienced environmental conservation measures in manufacturing plant in their own countries. Those companies operating in Vietnam are internationally well known so their brand names are recognised as product names in Vietnam. For such companies, any environmental damage caused by them could harm the reputation of their international brand images (Dao et al., 2011; Japanese

Ministry of the Environment, 2002).

6.2 Overview of Vietnam

Vietnam, located in the eastern part of Indochina, has a population of 84 million, the second largest country in Southeast Asia after Indonesia. Half of the population is under the age of 30 years. The capital, Hanoi, is located in the north while Ho Chi Minh City is the central economic city in the south. Since around 1989, when the reform (so called 'Doi Moi') policy started to have effect, Vietnam has achieved stable, high economic growth through encouraging investment by foreign countries and promoting industrialization. In 1995, the rapid economic growth in neighbouring Southeast Asian countries helped to attract lots of foreign capital into the country, leading to high economic growth. However, the boom was short-lived. After peaking at 9.5 per cent in 1995, the economic growth gradually slowed down each year to 4.8 per cent in 1999, mainly due to the influences of the currency and economic crisis in Asia in 1997 and the delay in creating a favourable investment climate. The Vietnamese Government responded to this situation by providing foreign companies with tax exemption and other incentives, and the growth rate rose back to 6.7 per cent in 2000, showing a recovery trend. Vietnam's economic growth rate has been among the highest in the world, in recent years, expanding annually at 7-8 per cent, while industrial production has been growing at around 14-15 per cent a year. The gross domestic product (GDP) growth rate for 2006 was 8.2 per cent. The entry into force of the US–Vietnam Bilateral Trade Agreement (BTA) in 2001 transformed the bilateral commercial relationship between the US and Vietnam and has greatly expanded business opportunities for American firms.

Vietnam became a member of the World Trade organisation (WTO) on January 11, 2007. Vast changes are expected in Vietnam's economy that could provide excellent opportunities for foreign businesses. To meet the obligations of WTO membership, Vietnam revised nearly all of its trade and investment laws and regulations. As a result, foreign investors and those seeking to sell goods and services to the increasingly affluent Vietnamese population benefit from the improved legislative framework and lower trade barriers. Local firms that have previously enjoyed a wide range of protection will now experience increased competition. International market pressures, customer requests for information, government's increasingly more stringent environmental regulations and policies are putting more pressure on firms to improve their environmental performance, of which, EIA and EMS are two most popular measures being implemented towards this end.

The development of the Vietnamese economy is centred on Ho Chi Minh City in the south, and Hanoi and Hai Phong in the north. Many companies are located in these areas and adjacent provinces of Dong Nai and Binh Duong in the south, and Vinh Phuc in the north, with more availability of estates established to cater for the increasing need of industrial production. For this reason, companies surveyed in this research were selected from these geographical locations.

6.3 EIA Practice in Vietnam

Since the emergence of EIA, there has been a growing interest in examining the effectiveness of this environmental management tool. In the 1990s, an international study on the effectiveness of environmental assessment highlighted several areas where improvements needed to be made. Scoping, evaluating significance, reviewing of environmental statements and post -decision monitoring and auditing that are often referred to as 'follow-up' were all identified as priority areas (Sadler, 1996). Lack of follow-up is arguably the weakest point in many jurisdictions. If effectively applied, an EIA should reduce any adverse environmental impacts of developments. However, without follow-up it only covers the predicted effects on the environment and the real effects are not fully considered.

Despite the recognised significance of EIA follow-up, the implementation is still limited. Barriers to implementation and elements of successful EIA follow-up, based on the experiences of practitioners from around the world (Morrison-Saunders et al., 2003; Arts and Nooteboom, 1999; Sebastiania et al., 2001), are presented in the following list.

Documentation

- Lack of documentation and document control
- Incomplete information
- Deficiencies in EISs

Communication

- Lack of internal communication
- Poor communication with neighbours

Environmental monitoring and measurements

- Lack of technical support
- Poor environmental treatment facilities

Legislation deficiencies

- Overlapping regulations
- Unambiguous regulations

Management commitment

- Lack of environmental awareness
- Resource limitations
- Limited support
- Lack of knowledge about benefits and need of impact assessment

The EIA system in Vietnam is implemented through Article 18 of the LEP and a series of implementing regulations, particularly Decree 175/CP and Decree 26/CP. Chapter III of Decree 175/CP contains requirements for the submission of EIAs by investors and enterprises, both foreign and local for appraisal. The result of the appraisal shall constitute one of the bases for competent authorities to approve the projects or authorise their implementation. Provisions prescribing the format and content of EIA reports are set out in the appendices to Decree 175/CP.

Chapter 3 of Decree 175 regulates EIA and defines which areas investors, project managers or directors of the offices and enterprises shall conduct an assessment of environmental impact. The scope for doing this includes assessing the current environment in the operating project area, assessing impacts of the project on the environment and proposing measures for environmental mitigation. If not empowered to a specific branch, it is the Ministry of Natural Resources and Environment (MONRE) that appraises the reports for the central level. At the local level appraisal is made by the provincial Departments of Natural Resources and Environment (DONREs). In case of necessity, an Appraising Council shall be set up and MONRE shall decide the establishment of the council. The chairmen of the People's Committees of the provinces and cities under the Central Government will decide the establishment of Appraising Councils at the provincial level. The time for appraising an EIA report would be within two months from the date all related documents are received.

The results of the appraisals over EIA reports are classified into four categories:

1. Being permitted to continue operations without environmental penalty
2. Having to invest in building facilities to deal with the waste materials
3. Having to change the technology, to move to another place
4. Having to suspend its operations

The Department of Appraisal and EIA attached to MONRE has been designated as the functional institution to assist the Minister in exercising the state management of EIA and appraisal. Guidance for EIA application in Vietnam is currently available in the General Guidelines Book (Le et al., 2000a). This applies to all major projects and covers many development sectors. However, these are complemented by sectoral guidelines that have been partly compiled by international bodies for their development projects. For example, EIA guidelines have been developed by the Asia Development Bank (ADB) on cement manufacturing and mining operations, thermal power generation and power transmission lines and highways; established by Vietnam-Canada Environment Project on hydropower dams projects; and those compiled by the 'Capacity Building for Environmental Management in Vietnam' project on tourism development and urban planning projects (Le et al., 2000b,c).

A number of different parties are involved in the management and implementation of EIA procedures in Vietnam including EIA managing agencies, proponents, environmental experts, other state management agencies, the public, international funding organisations, and universities and research institutes. Public participation is still a new issue in Vietnam. It is designated in the General Guidelines for EIA 2000 that public participation has legal importance and contributes to the success of the project but has not yet been practised regularly and nationwide.

The development of EIA in Vietnam can generally be divided into 3 phases since the 1980s. In the first phase which lasted from 1983 to 1993 preparations for EIA procedures were implemented. The activities undertaken during this period of time included training of EIA experts, elaboration of regulatory documents with respect to EIA, and adaptation of EIA methodologies into the current Vietnamese practical situation. The second phase from 1993 to mid 1996 involved the implementation of EIA procedures with the issuance of EIA related regulations, continuation of training of EIA experts, and elaboration and appraisal of EIA reports took place. The last phase from mid 1996 onward comprised the improvement of EIA expertise in regulatory and methodological aspects (Le, 1997).

The National Environmental Agency under the former MONRE has delivered guidelines for setting up EIA reports for different sectors, such as industrial park development projects and transportation projects. The sectors not covered still have to comply with the same regulations for content of the EIA report. MONRE is preparing guidelines to cover basically all sectors. The guidelines give comprehensive recommendations on the preparation of the EIA report including structure of the report, project description, proposed implementation of mitigation measures and so on.

MONRE has a department of inspection, namely the Vietnamese Environmental Protection Agency (VEPA). VEPA has a division of inspection and there are also inspection divisions under the provincial DONREs.

The current EIA procedure in Vietnam is basically consistent with the international practice that is implemented both regionally and worldwide. The EIA procedure in Vietnam can generally be categorised into four main steps as follow:

1. Screening
2. Preparation and submission of a form typed document, 'Registration for securing environmental standards', for projects classified in category 2
3. Preparation of detailed EIA report for projects classified in category 1
4. Appraisal of EIA report

All investment projects in Vietnam are required to be environmentally screened. Projects possess characteristics as delineated in Annex of Circular 409 will be classified in category 1 (Tran et al., 2000). These projects may contain apparent potential to induce adverse environmental impact, for instance, projects in or adjacent to environmental sensitive areas, oil and gas projects, etc. Thus, EIA is essential for projects classified in this category. Other projects will then be classified in category 2 where EIA implementation is not mandatory.

The screening procedure that conforms to the project classification is a way to facilitate the EIA implementation for both time and cost reduction.

The final step for projects classified in category 2 requires the proponent to prepare and submit the 'Registration for Securing Environmental Standards', to the environment management agency for appraisal (Le et al., 2000a). For a project classified in category 1 the preparation of a preliminary EIA report will be required before the subsequent EIA procedures. Detailed EIA will be initiated after the authorised bodies approve the preliminary EIA report. The appraisal of EIA reports will be conducted at different levels, including local, central or National Assembly, depending on the scale of the project.

The EIA procedure in Vietnam principally focuses on the establishment and appraisal of the detailed EIA report. However, about 70 per cent of the examined EIA-reports have major imperfections that need improvement (Tran et al., 2000). Most of these detailed EIA reports imply that practices such as impact identification, prediction, impact analysis, impact significance evaluation, and impact monitoring and management plan are not regularly integrated into the Vietnamese EIA procedures (Le et al., 2000a). Meanwhile the mandatory scoping in the pre-feasibility study used to prepare the TOR (Terms of Reference), and the impact monitoring and management plan that is implemented at the stage of post-construction are also not implemented extensively. The predetermined mitigation measures will therefore be non-applicable without a comprehensive environmental impact management plan at the operation stage of a project. In view of this, the overall EIA procedure has been reviewed by the relevant environment management agencies with the assistance from international funding organisations to establish a general guidelines book for EIA practice to provide a basis for upgrading the current EIA framework (Le et al., 2000a).

The integration of the EIA in the stages of feasibility study and decision making remains generally low. This may arise from the insufficiency of expertise needed to provide training and to carry out activities with respect to EIA, and the low integration between governmental environment managing agencies with other responsible stakeholders.

Meanwhile, decision makers and the general public find EIA reports difficult to understand because of their length and the complexity of the methods used. Apart from the managerial capacity, the technical capacity for preparing EIA's is also generally low. The scientific evaluation of EIA reports is still not a current practice in Vietnam, which subsequently raises another shortcoming that will reduce the opportunity to guide quality improvement. Therefore, there are some capacity building initiatives established by the cooperation between EIA managing agencies and international donor community in order to promote both the mentioned capacities for EIA practice (Le, 2010).

Project classification has simplified the environmental screening process. However, the application of the list of projects necessary to carry out EIA is questionable. In Vietnam, international and national nature conservation, historical and cultural heritage have already been defined. However, there are no criteria available to identify environmentally sensitive areas (UNDP, 1995). Thus, it is difficult to implement part 1, annex 1 of the circular No. 490 (Luc and Le, 2000). No specific framework is currently available to identify the potential adverse impacts that arise from the project scale or the ecological sensitivity for projects classified in category 2. This shortcoming may subsequently increase the risk of ignoring the potential cumulative impacts of the project. Therefore, it is necessary to establish more comprehensive sectoral and technical guidelines, which are still in their infancy in Vietnam. In view of this imperfection, a project collaborated with the EU, namely 'Capacity Building for Environmental Management in Vietnam' was conducted that aimed at the development of general EIA guidelines and specific sectoral EIA guidelines for hydropower dams, urban planning and tourism development projects (Le et al., 2000b,c).

Mitigation measures recommended in the EIA report and environmental impact monitoring are still not effectively implemented. As a consequence, most of the EIA reports prepared in this Vietnam lose their respective practical effects (Tran et al., 2000). As reported by Luc and Le (2000), firms commonly pay insufficient attention to the environmental management during the post-construction phase particularly the monitoring and auditing activities. This is evident from the lack of monitoring data conducted for comparison purposes with the predictive impacts in the EIA report. In view of this constraint, environmental impact management must be carried out by firms to conform to EIA procedures in the newly developed general EIA guidelines book. As emphasised in this document, the EIA report should be easy to understand and suitable for decision making, including qualifications, reliability and limits of the predictions on the environmental impacts (Le et al., 2000a).

There is a realisation that an EIA should be applied to all development project activities that will potentially cause significant adverse impacts or cumulative effects to the environment and the society. It should be carried out throughout the project cycle, start as early as possible, probably in the concept design stage. The contents of the EIA report should not only concentrate on development projects but also be elaborated to development plans at national, regional, and sectoral levels (Luc and Le, 2000). Of all the shortcomings of EIA practice in Vietnam, EIA follow-up is considered a big problem to be solved for EIA to meet its intended purposes. This necessitates the search for an environmental tool that can help to carry out EIA proposed mitigation measures and other related requirements specified in the EIA reports.

6.4 Environmental Management in Vietnam

In 1989, the Government of Vietnam embarked on a reform named 'Doi Moi' to guide the country from a centrally planned toward an open market economy. Vietnam has since opened up its economy and has entered the process of trade liberalisation. While this has been successful in generating strong economic growth it has created threats to the country's environment. In order to maximise the potential positive impacts and minimise the negative effects it is necessary to take measures to protect the environment. The ongoing reform process creates a good opportunity for the development and introduction of effective environmental management policies and instruments (Tran et al., 1996).

A number of measures have been proposed by the Government. In 1986, a draft 'National Conservation Strategy' was prepared. In 1991, a 'National Plan for Environment and Sustainable Development in Vietnam for the 1991–2000 period' was formulated by the former National Committee of Science, renamed to the Ministry of Science, Technology and Environment (MOSTE), and now to Ministry of Natural Resources and Environment (MONRE), with assistance provided by UNDP, UNEP, and the Swedish International Development Cooperation Agency (SIDA), and formally approved by the Government.

At the company level, although most companies have included environmental management in their operations to some extent, they have never had it inserted in the overall management framework. Environmental management is still not considered to be an issue which needs to be dealt with either strategically or systematically. There is some degree of consensus that the environmental management systems and ISO 14001 can be useful in improving the environmental performance of companies that assists in facilitating trade. However, it is recognised there is little experience and empirical research on the implementation of ISO 14001. The level of awareness of ISO 14001 reflected through number of certification and its benefits among businesses in Vietnam remains low compared to other countries in the regions, especially among the local enterprises (Japanese Ministry of the Environment, 2002; Vietnam Fatherland Front, 2008).

In Vietnam international companies are more concerned with environmental issues than local ones. Joint ventures and companies owned outright by foreign investors (100 per cent foreign invested companies) indicated a strong interest and conformance to ISO standards. State-owned companies in general, while accounting for about half the nation's mining and manufacturing production, execute almost no environmental conservation measures. For those domestic firms, environmental awareness is very new but is growing and will be an important tool for prevention of pollution by industry. Manufacturing enterprises are more concerned about the environment than other sectors. This is reflected in larger numbers of manufacturing enterprises implementing environmental management measures and undertaking other voluntary initiatives – for example, Toyota Vietnam Environmental Activities Grant Program. By 2005, 92 per cent certified companies are from the manufacturing sector, the rest are operating in the service sector. Many companies have executed firm environmental conservation measures based on the principle that these constitute a normal corporate activity. This is partly because of the pressure from their parent companies who promote environmental conservation measures in other countries. In addition, quite a few companies recognised a reduction of energy costs and production costs through the implementation of environmental conservation measures. By 2005, there were about 120 industrial estates and export processing zones in Vietnam. Some industrial estates, especially those managed by foreign investors like Japanese and Singaporean corporations, though constituting only a small fraction of these establishments, exercise excellent environmental conservation measures, thereby contributing to upgrading environmental conservation measures of Vietnam. These Japanese industrial estates naturally have their own environmental facilities such as wastewater treatment facilities. A Japanese company was the first in Vietnam to acquire certification of ISO 14001. By the end of 2005, more than 100 establishments in Vietnam had reportedly been certified, of which many were Japanese companies. The certified foreign companies used the process to enhance the environmental awareness of the Vietnamese senior members and operators by delegating to the Vietnamese staff and employees. Staffs participate in environmental conferences of companies from Southeast Asian countries, or even in environment related conferences in the parent companies' head offices in Japan and Singapore.

The gradual introduction of environmental legislation to regulate impacts on the environment has mostly involved determining compliance levels for pollution emissions but other environmental management tools have been developed. These include environmental auditing, environmental accounting, environmental reporting, life-cycle assessment, environmental management systems, risk assessment and environmental impact assessment. There may be a considerable difference between impact prediction and the actual occurring environmental consequences. The predicted effects are needed but it is the real effects that directly affect the environment. Follow-up is necessary to provide information about the environmental consequences of business activities as they occur, and also gives the responsible parties (proponent and/or competent authorities) the opportunity to take adequate measures to mitigate or prevent negative effects on the environment (Sadler, 1996; Arts et al., 2001; Marshall and Morrison-Saunders, 2003).

6.5 Determinants of Compliance Behaviour

From an understanding of determinants of firm compliance behaviour derived through the study, recommendations for policy making regarding environmental have been developed. The findings lend support to the framework of Scott's 'Three Pillars of Institutions' (Scott, 1995) encompassing regulative, normative and cognitive elements which together form the basis of firms' compliance behaviour. The elements vary among themselves and over time as to which elements are dominant. It also supports the use of triangulation approach to study the firm compliance behaviour.

The compliance norm is ranked first, followed by social influence, morality and the deterrence measures of regulators. Results from the interviews and the case studies also confirmed the importance of these attributes. Further analysis looking at the rating difference among groups of different size firms, fields of operation, business structures and ISO 14001 certified and non-certified firms found that those groups firms put a different level of importance on different reasons for compliance and non-compliance with different types of environmental requirements.

Firm compliance behaviour is found to be based on three institutional elements of organisational behaviour i.e. regulative, normative and cognitive. Each element reflects relevant business concerns: regulative elements include rules, laws, sanctions, violation detection and conviction and operational gains/losses; normative elements influencing organisational behaviour comprise of social influence, legitimacy, morality, capacity and commitment; the cognitive element is represented by firms' shared logics of actions.

Regarding regulative motivations, firms stated this as the fear of being detected, convicted and sanctioned for environmental violations. The probability of violation detection and level of sanctions are important factors which receive high attention from firms. The normative and cognitive elements are found to be important in determining firms' compliance with environmental laws and regulations: businesses are likely to implement environmental programmes under the pressure from public forces like stakeholders, community and peer groups and their perception of the legitimacy of regulations and regulators. They are very concerned that their social reputation would be enhanced by good environmental performance and adversely be harmed by poor performance in environmental management activities. The appropriateness and effectiveness of laws and the fairness of regulators are also important in determining firms' compliance with the regulations. The improvement of workers' health, safety and welfare and environmental protection are important for firms determining their implementation of environmental management activities.

Firms implementing voluntary programmes like ISO 14001 EMS are also driven by the economic benefits of a system that helps to reduce their operating costs, material wastage and to enhance the firms productivity. This, on the other hand, seems not to be the perceived benefits of regulatory measures. This lends support to the indication that regulatory measures, like EIA, are more of a paper exercise to gain project approval rather than having any practical benefit for firms.

Firms operating in the field of manufacturing and construction are more concerned about implementation cost, community pressures and the environmental protection outcomes of different environmental initiatives. Service sector companies in contrast put more emphasis on international market access as important reasons driving their implementation of environmental management measures. The results reflect the fact that manufacturing and construction sector have more negative impacts on the environment than the service sector and thus are under more pressure, from the public and the government, to take appropriate measures to minimise their impacts. Many sampled industrial firms have foreign businesses operating in Vietnam with a focus on the local market while service firms are all in the hospitality area serving the international market. A difference was also shown in the concern for social reputation between state and private sector enterprises. This seemed to reflect an uneven playing field between these two economic sectors in the Vietnamese market in which private firms have many disadvantages over their competitors in terms of incentives and subsidies from the government. Enhancing their social reputation is one of the efforts the private sector enterprises make to compete with state companies.

Firms cited three important reasons for non-compliance with both regulatory and voluntary requirements. The reasons received different rankings but are generally considered to be among the seven most important reasons for non-compliance. The identified reasons reflect the logic of appropriateness that firms follow when implementing environmental management programmes including agency losses, ignorance of law/difficulties in understanding environmental regulations, lack of financial and technological ability to comply.

Capacity, commitment and social influence are important reasons determining compliance. The complexity and ambiguity of rules and regulations and the enforcement problems often lead to the ignorance of, or difficulties in understanding relevant environmental regulations for businesses, resulting in non-compliance. Low environmental awareness of workers and coordination problems within firms may result in agency losses with employees disobeying owners' and staff management orders to comply. Firms with a low level of financial and technological ability also have problems exhibiting good environmental performance. This is supported by the additional comments by responding firms regarding the lack of treatment facilities and low management capacity of environmental agencies, especially in terms of waste water treatment and solid waste and hazardous waste collection and treatment.

The implementation of EIA encounters certain obstacles that are considered important to this environmental management tool but not other environmental management programmes under study. Due to the nature of EIA as a pre-decision making procedure and its implementation having been applied at an early stage of project implementation, 'low management awareness' and 'lack of leadership concerns and commitment' do not seem to be important reasons hindering EIA's implementation. On the other hand, the highly technical and complex nature of EIA make cooperation of local government very important for the implementation of EIA. Lack of leadership concern and commitment is considered one of the most important reasons hindering the success of ISO 14001 EMS implementation and certification. Besides, the implementation of this voluntary environmental management system also encounters difficulties in changing working traditions of the employees, an attribute not important for the implementation of other environmental requirements. This finding is supported by the findings on the difference rating between ISO 14001 certified and non-certified firms: ISO 14001 certified firms are highly aware of the difficulties in changing working traditions of employees and the difficulties made by lack of leadership concern and commitment in the implementation of this environmental management system.

Small firms, mostly private ones, are more concerned about legitimacy of laws, increased operation cost and the availability of human resources to comply than large firms. Smaller businesses usually have less available resources and time to address environmental issues.

It is noted from the findings that state owned enterprises are not very concerned about financial issues of the implementation of environmental management measures. The state owned companies have easy access to financing for environmental management while other forms of businesses have to be very efficient in every investment decision (Tenev et al., 2003).

6.6 The Role of ISO 14001 EMS in EIA Follow-up

The interview and survey results showed that the implementation of an ISO 14001 EMS can cover the requirements of EIA follow-up. Within the scope of ISO 14001 EMS, more impacts are identified than predicted in EIA and more management activities are actually implemented compared to the number of mitigation measures proposed in EIA, to address the wider range of identified impacts. These findings are confirmed by all three sets of data of case studies, interviews and survey.

EIA is regarded as an important reference document during the implementation of ISO 14001 EMS, providing necessary information on the relevant environmental legal requirements of the project and supporting data on project's baseline conditions. The finding of more impacts being addressed by an EMS than predicted in an EIA report is explained by the higher level of concern of companies on environmental issues at the operational stage of the project leading to a company's higher environmental targets and objectives. The tighter requirements of the ISO system, legal updates, practical nature of EMS, production expansion and pressure from customers are also relevant.

In addition, some EIA predicted impacts do not actually occur in practice. There are several reasons for their non-occurrence: some proposed mitigations measures may have been applied to help to minimise the impacts on the environment; other explanations include inaccurate predictions, removal of production facilities and activities and change of production plans.

In summary, the implementation of management activities under ISO 14001 EMS have been shown to meet or even go beyond EIA follow-up requirements. EIA follow-up is the minimum legal requirement that firms have to comply with and compliance is also within the scope and environmental objectives and target of ISO 14001 EMS. The findings of the study show that it is possible to improve compliance with environmental policies and other voluntary programmes through a combination of measures that address the wide range of institutional elements affecting organisational behaviour.

7. EIA–EMS Link from the Oil and Gas Industry

Behzad Raissiyan and Jenny Pope

7.1 Introduction

The oil and gas sector has something of an image problem when it comes to environmental and sustainability performance, not helped by high profile disasters and controversies such as Exxon Valdez (Exxon), Brent Spar (Shell), and more recently Deepwater Horizon (BP) that are well established in the global public consciousness. Other incidents with catastrophic environmental and social outcomes receive less attention in the international media: for example there were 17 catastrophic incidents in the Iranian oil, gas and petrochemical industries during 17 months from March 2010 up until July 2011, including large oil spill cases, gas leakages, and major fires, with about 30 fatalities and more than 100 major injuries (Khabaronline, 2011; Ayande Independent Iranians Online Press Media, 2011). Coupled with growing concerns about climate change and the contribution of fossil fuels to carbon dioxide levels in the atmosphere, it is clear that despite significant efforts within the industry to improve environmental and social management and performance in recent decades,[1] the sector as a whole is a long way from being considered sustainable – for example, see Harden and Walker (2001); Ugochukwu and Ertel (2008); Harris and Khare (2002); Moser (2001); May (2003).

However, given that fossil fuels are a fundamental component of industrial economies, and that even the most optimistic scenarios of transition to a greener

[1]For example through the efforts of international industry organisations such as the International Petroleum Industry Environmental Conservation Association (IPIECA), International Association of Oil and Gas Producers (OGP), and Energy and Biodiversity Initiatives (EBI).

economy cannot ignore the crucial role of fossil fuels, at least in the short and medium terms, the immediate imperative is to continue to improve the environmental and sustainability performance of the oil and gas sector. This chapter aims to contribute to this improvement in one very important aspect of this greater challenge: improving the links between pre-decision (or pre-implementation) activities such as environmental impact assessment (EIA) and post-decision (or post-implementation) efforts such as environmental management systems (EMS) for oil and gas projects. We argue that while considerable efforts and investments are made in both EIA and EMS for a typical project, a lack of integration of these activities contributes to sub-optimal environmental outcomes and a failure to realise the potential of these well-established environmental management tools. While this statement is undoubtedly true of many sectors (Marshall, 2002; Varnäs et al., 2009), we believe that the complexity and sheer size of major oil and gas projects exacerbate the challenges faced by environmental practitioners assigned to these projects.

This chapter presents some practical initiatives that have been taken in two major oil field development projects in Iran, in order to strengthen the linkage between EIA and EMS throughout the life cycle of these projects. Our analysis and subsequent recommendations are based upon the personal experiences of one of the authors who worked on these projects, in one case as an EIA consultant and in the second as an internal EMS practitioner. Arguably the challenges associated with effective EIA implementation and follow-up are even greater in developing countries such as Iran, where lack of capacity and inadequate regulation are common (Arts et al., 2001; Jha-Thakur et al., 2009; Ahmadvand et al., 2009). Despite the apparent barriers, however, we argue that there is much that proactive and determined environmental practitioners can achieve within the project team, and that implementing effective EIA follow-up is one of the most significant opportunities for improvement in environmental management in this sector.

In Section 7.2 we present some of the general challenges and barriers to the integration of EIA and EMS that we perceive based upon our experience of oil and gas development projects. The two case studies demonstrating practical actions to improve the integration between EIA and EMS in oil development projects in Iran are presented in Section 7.3, and finally overall conclusions and recommendations for the way forward are discussed in Section 7.4.

7.2 Environmental Management in Oil and Gas Development Projects – Challenges and Issues

While this chapter focuses on the challenge of ensuring strong linkages between EIA and EMS in the context of an oil and gas project, and seeks to demonstrate how these environmental management tools can be more effectively integrated through project delivery to deliver effective EIA follow-up and good environmental performance, it is worth noting two other important factors that often contribute to sub-optimal environmental outcomes in this sector:

Inadequately defined strategic or policy context The principle of 'tiering' in impact assessment suggests that strategic levels of decision making, including policies, plans and programmes to which strategic environmental assessment should ideally apply, should provide a framework for project level decision making (Noble, 2002; Thérivel and Partidário, 1996). An oil and gas development project, for example, might be located in accordance with a regional development or land use plan, perhaps to meet policy objectives described in a sectoral strategy, such as a national energy plan (Sadler, 1996). However, it is all too often the case, particularly in developing countries and those with a high level of economic dependence on oil and gas, that these strategic levels of decision making are either not conducted at all, are out of date, or do not adequately reflect environmental and sustainability considerations, leaving the project to be developed in something of a policy vacuum and increasing the chances of multiple projects being developed in sensitive areas with significant cumulative environmental impacts.

Lack of EIA during the exploration phase Oil and gas exploration activities, including seismic surveys and exploration drilling, can themselves generate significant environmental impacts. In many countries, however, there are no requirements for EIA of these activities, since at this early stage the potential investment is not yet defined as a project. In Iran, for example, exploration activities are merely considered surveys under the jurisdiction of the Ministry of Petroleum.

These characteristics of current practice establish a challenging context within which the EIA and EMS practitioners on a project must operate, since they place limitations on the effectiveness of environmental management founded in project EIA and EMS. The main purpose of this chapter, however, is to explore how environmental issues are managed during the life cycle of a project from the point it which it can be considered a project, so we will not devote further attention to these strategic and upstream issues here.

Oil and gas projects are usually very large and they are inherently complex. Due to the vast investments they require, projects are planned and designed progressively in layers of increasing detail, and financial viability is reviewed many times until the Final Investment Decision (FID) is made. Typical stages of this process up to FID include Pre-Feasibility, Feasibility, Conceptual Design, Pre-Front End Engineering Design (pre-FEED) and FEED, followed by Detailed Engineering Design, Construction, and Commissioning. EIA is conducted in accordance with legislated requirements (now applicable in most countries of the world) for the purpose of determining whether a proposed project is environmentally acceptable and for establishing the environmental performance conditions under which it is allowed to proceed.[2] The impact mitigation strategies identified in the EIA, which usually become conditions of approval, must then be managed throughout the project life cycle through the development and implementation of an EMS.

How and when the transition between these two key environmental management tools occurs within the long and complicated project life cycle varies in oil and gas project practice. We argue that ideally, EIA should be considered as a flexible process which starts at the pre-feasibility stage of the project, continues through pre-decision activities and the point of decision into the post-decision activities, where it should merge seamlessly into the project EMS as the project is developed and implemented, in a way that challenges the very notions of 'pre-decision' and 'post-decision' (Morrison-Saunders et al., 2007).

In our experience, however, this rarely occurs and there is commonly a failure of integration and alignment between EIA and EMS, which results in the

[2]In many cases social impacts are also considered as part of an EIA through a parallel social impact assessment (SIA) process. We do not address the social dimension in this chapter but the principles we discuss are equally relevant to the management of both social and environmental issues.

commitments made through the EIA process not being adequately implemented through project delivery. We believe that the following characteristics of major oil and gas projects, and probably other large resource projects, can be attributed to factors including the following:

Timing of the EIA It is often argued that an EIA should be conducted 'prior to major decisions being taken and commitments made' (IAIA, 2009), or at as early a stage as reasonably possible (Duffy and DuBois, 1999; Sadler, 1996). In practice, however, there is a tension between conducting the EIA too early, when there is insufficient project information available to enable meaningful evaluation of impacts (Walmsley and Tarr, 2011) and conducting the EIA too late, when agreements with the key contractor(s) are in place (Walmsley and Tarr, 2011) and/or the project design is so close to being finalised there is little flexibility to influence the project design and/or choose between alternatives (Duffy and DuBois, 1999).

Insufficiently specific mitigations It also appears to be common that mitigations proposed in EIA reports are often very generic and fuzzy, a problem that is exacerbated if the EIA is undertaken too early and one noted by Sánchez and Hacking (2002). Authorised agencies, decision makers and stakeholders in many cases demand very detailed information about likely associated impacts and also very detailed and concrete mitigations which only can be provided after FEED (and in some cases after Detailed Engineering Design). In many instances, as discussed further in the context of our case studies, the proposed mitigations are also too generic to be understood and effectively implemented by EMS practitioners.

Different languages, methodologies and perspectives While most EIA practitioners have a background in ecology, environmental science, biology, or social science and may lack industry experience, EMS practitioners typically have a background in engineering disciplines with considerable practical industry experience. It is rare for practitioners to be experienced in both EIA and EMS (Marshall, 2004). The resulting different perspectives and different terminologies, which are also reflected in EIA and EMS literatures – as noted by Sánchez and Hacking (2002); Marshall (2004) – can create additional challenges to the integration of EIA and EMS.

Multiple players The project owner who commissions the EIA report and obtains the permit is clearly ultimately responsible for EIA implementation. Typically, however, after decision making and permitting is complete, the main role player is the Contractor (typically a large oil and gas Exploration and Production Company), who is responsible for the project EMS and often perceived by the community and stakeholders to be the project proponent. There is therefore a 'baton change' between the project owner and the EIA, and the Contractor and the EMS – see also Marshall (2004). The disconnect is exacerbated when the project owner itself enters into contracts with some other contractors/sub-contractors to complete some parts of the job. The responsibilities for EIA and EMS implementation are therefore shared and it is common for roles, responsibilities, authorities and accountabilities for environmental issues to be poorly defined and communicated.

In the following section we present two recent case studies that demonstrate practical ways in which EIA and EMS can be better integrated in large-scale oil and gas projects, lessons that we believe are likely to be transferable to other sectors.

7.3 How Practitioners can Enhance the Integration of EIA and EMS

In this section we present some practical examples of strengthening the linkages between EIA and EMS in two major oil projects. The oil and gas sector is usually divided into two sub-sectors: 'upstream' and 'downstream' (E&P Forum, 1997). The project case studies presented here are both examples of upstream projects, which typically involve a broad range of activities including exploration, drilling, construction, piping, production (oil/gas dehydration, separation, desalination, treatment etc.), together with supporting services activities. On the other hand, downstream oil and gas processes may include oil refining or gas processing.

The two case studies are similar in many ways: both projects involved oil field development; both were located in the west of Iran, in the border area between Iran and Iraq which was severely affected by the eight years of conflict between Iran and Iraq (1980–1988); and both were characterized by similar environmental and socio-economic conditions. In summary, both areas were covered by very poor and low density steppes (semi desert) which were already significantly deteriorated by over grazing. Remnant Mesopotamian marshes (registered wetlands) were located in the project vicinity of the second case study. There were no permanent settlements around the project area in either case: there were only some migrant shepherds who migrate seasonally to the area looking for fresh pastures for their livestock. There were still some land mines and unexploded ordinance (UXO) remaining from the war.

In the first case study project, the primary author of this chapter was in the position of the EIA team leader and senior assessor. This case study therefore focuses on the ways that EIA practitioners can influence EIA implementation. In the second case study project this writer was in the position of EMS practitioner (EMS advisor for the project). This situation provided a unique opportunity to apply the lessons learned from the first project in the second, and to explore the challenges of EIA–EMS integration from a different perspective. While the circumstances were somewhat different for each project, key to the success of both case studies was the effective deployment of EIA outcomes throughout all phases of project delivery and the engagement of all project teams and disciplines.

7.3.1 Case Study 1: How EIA Practitioners Can Support the Integration of EIA and EMS

This project involved the renewal and restoration of an oil field that had been initially developed about 40 years ago, but the majority of whose facilities, wells and pipelines had been extensively damaged during the Iran-Iraq war. Following some minor repairs after the war, in 2008 the Iranian Ministry of Petroleum approved a plan to renew and restore some old and/or damaged on-shore oil fields. One of the subsidiaries of the National Iranian Oil Company was assigned to lead these projects. The oil field itself consists of six different oil/gas reservoirs, divided into two operational areas (ICOFC-West, 2009).

The original development (or more accurately renewal/restoration) project was divided into several sub-projects. The Terms of Reference (TOR) for the EIA correspondingly described four separate EIA packages, and the EIA teams were requested to ensure that likely interactions, cumulative impacts and overlaps between all packages were addressed and taken into account sufficiently. The case study analysis that follows is based upon experiences with two of these packages that were awarded to the consultancy firm represented by the primary author, and which can be considered as one integrated package:

Package 1 (2008) Development and construction of a new Central Natural Gas separation and treatment plant and associated export pipelines; and

Package 2 (2009) Drilling and/or work over and completion of 16 oil wells with a capacity of 55,000 barrels of crude oil per day together with associated infrastructure

The EIA studies for these two packages were conducted in parallel with the Conceptual Design stage of the project, which provided some opportunities for the EIA process to influence certain aspects of project design and development, as discussed further below. The importance of working closely with the technical teams was also recognised in the first instance by the EIA team, whose tender documentation included an option of 'full interaction with the project design team'. Here it was proposed that a multi-disciplinary advisory team including a drilling/reservoir advisor, process/production advisor, construction advisor and surface facility (mechanical/process engineer) advisor would accompany the EIA team and participate in all communication meetings and workshops. It was also proposed that a representative from the EIA team would attend all relevant project meetings.

Unfortunately, the client chose not to accept this proposal in the interests of minimising costs, but the EIA team was still able to find ways to work with the project team and to facilitate the integration of EIA outcomes into project planning and design activities. The relationships with the teams responsible for the Quality Management System (QMS) and other aspects of the integrated Health, Safety and Environment Management System (HSE–MS) were particularly important. Some of the key successful initiatives are outlined below.

Interaction with the project design team

Despite the client's decision not to accept the full integration package as proposed, a number of successful interactions between the EIA team and the project team were initiated, including:

- Including as a key quality objective for the EIA the commitment to add value to the project design and implementation, and to the project QMS and HSE–MS through effective communication and interaction with project technical teams
- Three interaction workshops conducted at the following milestones:
  - Evaluation of impacts (including identification, prediction and significance assessment)
  - Development of proposals for mitigations and enhancements and evaluation of residual impacts
  - Finalisation of mitigations and development of the Environmental Monitoring Plan
- A joint site visit by the EIA team and project team

At the end of the EIA study, in accordance with the EIA Quality Plan, client and stakeholder satisfaction was evaluated through interviews and questionnaires. The results revealed that the most satisfying aspect of the EIA process was the interaction workshops and the positive outcomes and added values that were achieved through them. This finding sits in sharp contrast with the conventional perception that clients are only concerned with EIA for the purposes of legitimising and obtaining approval for their projects (Marshall, 2002). Some of the interviewees even indicated that had they been aware of the importance of proposed interaction programme, they would have supported and approved it during the tendering process. Some of the key outcomes of the implementation workshops were:

Site/route selection Since the EIA study was conducted in parallel with the conceptual design the EIA team had an opportunity to ensure that environmental and social criteria were included in the GIS-based multi-criteria analysis process that was applied to identify optimal sites and pipeline routes.

Mitigation measures All mitigations proposed by the EIA team were checked with the project design and implementation teams in the second workshop. Many of the proposed mitigations, which were initially perceived as being ‘too generic’ by the technical teams, were successfully translated to technically meaningful inputs to design and implementation (see Example 7.1).

Inputs to QMS and HSE–MS The interaction workshops also helped to ensure that EIA outcomes were prioritised along with other project risks; categorised as either substantial (physical) actions (those relating to changes in project design for example) or procedural actions, and effectively incorporated into the project QMS and HSE–MS as well as into the tender documentation for subsequent contracts. To facilitate compliance monitoring, a Central Environmental Compliance Database was developed and handed over to the HSE management team as described below.

Example 7.1 Example of an effective multi-disciplinary approach to developing EIA mitigations

The generic action of 'Avoiding burning of produced crudes associated with well testing' was investigated by both the EIA and technical teams. One option that was considered involved using a pipeline to send produced hydrocarbon to the existing manifold. At first, this seemed to be unacceptable due to the considerable length of temporary piping required for only one day of well testing. But further investigations revealed a very different result. In fact the project planning and control team suggested that if they changed the project implementation schedule to bring the piping forward to well testing, it would be possible to use permanent pipelines for this purpose. Cost engineers and other disciplines all confirmed that this alternative was workable, even though it required major changes to the contract strategy and the Project Execution Plan (PEP). The result was not only much more environmentally sound, but also much more time and cost effective. In this case the key success factor was the involvement of cost engineers and project planning and control team in developing the EIA mitigation; without their participation this outcome could not have been achieved.

Central Environmental Compliance Database

A comprehensive database of all environmental and social requirements was developed by the EIA team and handed over to the client's environmental management team. The term 'requirements' in EMS generally refers to all legislation and regulations, conventions, protocols, relevant industry specific codes and standards, internal company/corporate requirements and/or any other imposed requirements by stakeholders (ISO, 2004). Compliance with applicable legal and other requirements is one of the three pillars of all HSE–MS Standards and guidelines.[3] EMS practitioners usually are more familiar with this terminology than with EIA language such as 'mitigation measures' or 'EIA outcomes'. This in itself is a real contributor to the disconnect between EIA and EMS, which this database helped to overcome by translating from the language of EIA to the language of EMS.

The HSE management team was assigned to be the database custodian and to keep it up-to-date, and to regularly evaluate compliance, the database was open for all disciplines to input their specific requirements and to access the environmental requirements applicable to their activities. Interviews and questionnaires revealed that the HSE management team of the project, as well as representatives of other disciplines, found this database very useful in conducting their activities and meeting their environmental responsibilities. More detailed information and some examples of this database and its role in EIA–EMS integration are presented in the analysis of the second case study.

Alignment of the EIA Outcomes with the Project HSE–MS

To make EIA outcomes more digestible and easier to follow for the HSE team, the proposed environmental management plan (EMP) that formed part of the EIA was developed using the language and structure of ISO 14001:2004 as a standard EMS/HSE–MS model. In addition, a list of required procedures, work instructions, and guidelines to be developed and implemented as the project proceeded was provided for the project HSE team, once again aligned with the project HSE–MS elements. Finally the project EMS Manual with the proposed environmental management structure was outlined, and the key requirements of

[3]Two others are continual improvement and prevention of pollution/impacts/accidents.

the EIA were incorporated into the project HSE policy, strategic objectives and goals.

7.3.2 Case Study 2: How EMS Practitioners Can Support the Integration of EIA and EMS

This case study relates to the first of three development phases of an oil and gas development, involving the drilling of 45 production wells with a target production rate of about 85,000 barrels per day (bbl/d), three appraisal and three produced water injection wells, two manifolds, wellhead facilities and crude flow lines (inside the oilfield), the Central Treatment and Export Plant which includes a gas and oil separation unit, permanent accommodation camp, power generation and utility units, water supply and waste water treatment units, and finally a 130-km 20-inch export pipeline. Phase 1 has been wholly assigned to an international upstream Exploration and Production Company (Contractor) under an Engineer, Procure, Construct and Commission (EPCC) contract.

The EIA for Phase 1 was undertaken by a local consultancy under contract with the project owner, commencing before the project feasibility study had been completed. In line with Iranian legislations, the EIA report was submitted to the Department of the Environment (DOE) and local environmental authorities and a permit was obtained. The project contract was signed and the Contractor commenced its activities after the permitting process. The project timing and contractual arrangements meant that the EIA team and the EMS team represented different organisations and had no interaction and therefore no opportunity to influence each others' processes, creating a chasm between the two environmental management tools.

In accordance with the terms of the contract, the Contractor was responsible for implementing a comprehensive HSE–MS in line with good international practices, of which the EMS was a component. As far as possible, the HSE–MS was also integrated with the QMS, in acknowledgment that environmental incidents that may have severe impacts on the receiving environment are almost always caused by technical failures rather than failure of environmental management. For example HSE considerations were explicitly integrated with management of change processes, contractor management procedures and asset management processes.

The integration of the EMS and the EIA, however, was more challenging. The only clearly articulated link between the EIA and EMS was indeed the 'compliance with requirements'; in other words, EIA outcomes were supposed to provide inputs to the EMS. However, the EMS team quickly realised that the mitigation strategies proposed in the EIA were overly generic and would have been difficult to implement due to the lack of detailed project information available to the EIA team at the early stage of project planning in which the EIA was conducted. Faced with these challenges, the EMS team implemented a number of initiatives to align and integrate the EIA and the EMS and to ultimately deliver good project environmental outcomes. These are discussed below.

Developing and communicating the Central Environmental Compliance Database

Similarly to the first case study, a comprehensive database of all applicable environmental requirements was developed, together with a guideline manual, and this was integrated with the broader EMS Information Management System for the project.[4] This Information Management System is a fundamental tool to support the alignment and integration of all EMS elements with EIA outcomes. The Compliance Database and its guideline manual incorporated: the identification of all applicable requirements from different sources, including the EIA; the deployment of EIA requirements throughout the project team by identifying specific actions, applicable project phases and responsibilities; and an action tracking and compliance database.

During Conceptual Design and FEED all requirements identified in the database were audited and supplementary actions proposed where applicable. Furthermore, as project design and development proceeded, different technical disciplines and design teams provided their input in formal compliance evaluation and interaction workshops, through which the initial generic mitigations were translated into agreed specific and practical actions. An example of how EIA outcomes were deployed and cascaded during project development is provided in Example 7.2.

[4]Other components of the EMS information management system included environmental indicators monitoring, non-conformities and incidents, hazards and environmental aspects and

Example 7.2 Example of deployment of EIA outcomes

In the EIA report, the importance of the remnant Mesopotamian marshlands in the vicinity of the oil field had been addressed, and 'minimising intervention and landscape fragmentation and compatibility of oil field development with future rehabilitation' had been proposed as a mitigation measure. During Conceptual Design and FEED, this generic requirement was converted to the following elements, and then incorporated into the Basis of Design documents (as mandatory requirements for the design team):

- Minimisation of the footprint of the project by clustering of drilling wells where technically possible and centralising of surface facilities to the maximum extent possible;
- Minimisation of the footprint of the camp, roads, pipelines and general infrastructure design compatible with potential future restoration of the marshes;
- Maximising usage of existing/already planned infrastructure, such as existing roads and export pipeline routes;
- Allowing for natural surface water flow, including measures such as culverts, considering potential future restoration of the marshes;
- Elevated surface facilities and drilling sites, and minimisation of potential for pollution of the low-lying marshland areas.

Implementing the Environmental Compliance Database: aligning risk assessment processes and terminology

The core of both EIA and EMS practice is the identification and prioritisation of environmental issues. In the terminology of EIA this involves the identification of impacts and evaluation of significance. In practice, impact identification is typically quite high level, reflecting the level of detailed project information available at the time of the EIA, and significance evaluation is qualitative, with

impacts, meetings, training sessions etc.

no real standardised approach. In contrast, EMS draws more from the disciplines of engineering and health and safety, and the key techniques upon which the entire EMS is founded are typically hazard identification and risk assessment. Hazard identification and risk assessment techniques, particularly in the oil and gas industry are highly sophisticated, largely standardised and in many cases quantitative. Common techniques include hazard and operability studies (HAZOP), hazard identification studies (HAZID), Job Hazard Analysis (JHA), and Failure Mode Effects Analysis (FMEA), which are applied at different stages of the project development in increasing levels of detail.

The challenge for the EMS team was therefore to ensure that the generic impacts identified in the EIA were effectively incorporated into the broader project hazard identification and risk assessment processes to enable increasingly specific and refined mitigation strategies to be developed as the project progressed. The dissemination of the Central Environmental Compliance Database throughout the project team was the mechanism that enabled the environmental impacts that were identified and evaluated for significance in the EIA to be continually reviewed, classified, evaluated and communicated as the basis for EMS/HSE–MS development and establishment. All relevant teams were required to complete, update and detail the aspects, hazards and impact lists and respective mitigations, barriers and operational control as the project progressed. This approach is consistent with the principles of best practice EIA follow-up which state that EIA should not be considered complete after the approval and permitting process, but should be conceptualised as a dynamic process to be continually deployed as the project proceeds (Morrison-Saunders et al., 2007).

The implementation of this process proved challenging in practice, however, and an internal audit revealed that some of the environmental requirements were simply ignored, others were not addressed adequately, and that fairly standard risk assessment processes were conducted that did not adequately address environmental concerns. Interviews with HAZID/HAZOP teams, conducted as part of a root cause analysis of the identified non-conformances, revealed views such as: 'the compliance database is difficult to understand and follow', in some cases it was 'irrelevant' or 'too generic', or 'not fit for the purpose', or 'technically meaningless'. This highlighted the language barriers that existed even between the environmental and the health and safety practitioners within the HSE–MS team, much less the EMS team and the greater project team.

This very critical finding led the EMS team to take a new approach of actively supporting the project teams in meeting their environmental requirements. The EMS team proactively participated in risk assessment workshops and also conducted some complementary workshops focusing specifically on environmental issues, rather than simply communicating a long list of high level requirements and expecting each team to comply with requirements. Risk assessment workshops usually commenced with a short briefing session, in which the EMS team explained the requirements and functions of the Central Environmental Compliance Database, and how it should be incorporated into the risk assessment process. Through the multi-disciplinary risk assessment workshops, the EMS team was able to communicate the EIA/EMS key messages to the whole project team. It was a golden opportunity for information exchange about the environmental issues, the provision of on the job training and the development of awareness of environmental commitments and responsibilities across the whole project. A tangible example of bridging the language barrier is provided in Example 7.3.

Environmental performance monitoring and reporting

Environmental monitoring requirements also differ between EIA and EMS. While EMS monitoring typically has an internal, process focus, with an emphasis on monitoring the key characteristics of operations and activities that could impact on significant environmental aspects (ISO, 2004, Clause 4.5.1), effective monitoring from an EIA implementation and follow-up perspective should also incorporate an external, outcomes-based component through which the actual impacts on the environment are measured. Effective environmental monitoring from an EIA perspective can be considered to include (Sadler, 1996; Lohani et al., 1997; Duffy and DuBois, 1999; Sadler, 1996; Canter, 1998; Arts et al., 2001):

- Baseline monitoring, to establish the ambient environmental conditions prior to project implementation as the basis for monitoring project impacts;
- Compliance monitoring (as emphasised in EMS), to monitor compliance with EIA conditions, which could relate to the implementation of required mitigation actions, or to the achievement of emissions targets for example,

Example 7.3 Example of aligning risk assessment processes and terminology

The well-known Bow-tie model had been used in the project to identify hazards and impacts associated with project activities, to analyse causes and effects, to assess the risks for each identified hazard, and to propose mitigations and barriers to control those hazards/risks. To evaluate the significance of risks, a Risk Assessment Matrix had been developed, which defined significance criteria for impacts on: people (in terms of injuries, fatality etc.), the corporate reputation, the project assets, and the environment. The defined criteria for assessment of consequences to the environment and to the corporate reputation were fuzzy, defined in terms such as 'minor impact', 'major impact' etc.). Members of the EMS team participating in the risk assessment sessions who were familiar with both industry risk assessment terminology and also EIA/EMS terminology, suggested more specific and accurate definitions for significance criteria that aligned directly with the EIA outcomes. Concepts of stakeholder involvement, cumulative impacts and adaptation were adopted from the EIA report and incorporated into the risk assessment criteria.

Furthermore, the proposed EIA mitigations were presented in the language of the risk assessment process as being 'existing or planned actions' for which the residual risk could be assessed. Thus, if the proposed mitigations were found to be inadequate to reduce risk to an acceptable level, additional mitigations and practical actions were then proposed by the team. In this way, environmental concerns were fully integrated into the broader risk assessment and management practices employed by the project team.

depending upon how the EIA mitigations are defined;

- Impact monitoring, which should include monitoring the conditions of the receiving environment to determine whether the project's impacts are within acceptable levels.

Baseline monitoring and impact monitoring need to be aligned in order to provide a basis for detecting cause/effect relationships between impact sources or 'stressors' (environmental aspects in EMS terminology) and changes of the state of the environment or 'receptors' (Beanlands and Duinker, 1984). This is the only way in which organisations, stakeholders and authorities can effectively monitor and manage the cumulative impacts of projects. To monitor environmental performance and provide a basis for benchmarking and monitoring of continual improvement, performance indicators need to be defined. Nowadays, most oil and gas companies have an environmental performance indicator profile, mostly aligned with OGP/IPIECA recommended indicator profile – see IPIECA/API/OGP (2005, 2010); OGP (2010). However, these tend to focus on operational performance and do not cover environmental condition monitoring, which is crucial to effective monitoring from an EIA perspective (Eccleston and Smythe, 2002). It is useful to note that the ISO14031:1999 Environmental Performance Evaluation Guidelines speak of 'environmental condition indicators' as well as 'environmental performance indicators', which in turn are divided into Operational Performance Indicators (OPI) and Management Performance Indicators (MPI).

Another key difference between EIA and EMS reporting requirements is that EMS/HSE–MS standards do not impose specific performance criteria, since acceptable levels of performance depend upon the industry and the environmental context within which it operates. Consequently, good environmental performance from an EMS perspective is usually defined in terms of compliance with 'legal and other requirements', benchmarking against industry standards, and continual improvement of environmental performance. In contrast, monitoring from an EIA perspective should also ensure that actual environmental impacts are acceptable, regardless of whether or not legal requirements or industry best practice emissions standards are being met. Acceptability will depend upon the nature and sensitivity of the receiving environment, which should be established through the EIA.

To ensure an effective environmental monitoring programme for this project, all monitoring requirements imposed by the EIA and EMS were incorporated into a single 'HSE Performance Monitoring and Reporting Procedure'. The indicators incorporated leading (proactive) and lagging (reactive) indictors, as is common practice for HSE monitoring in the oil and gas industry, and also distinguishes between core, additional and supplementary indicators for which monitoring and reporting requirements are different. For example, the volume of flared gas/burnt hydrocarbons was included as a core lagging indicator. The number of visitors to the environmental pages of the website was included as a supplementary leading indicator.

The EIA imposed monitoring requirements are completely integrated with the other environmental monitoring schemes. As an example, the EIA report requires minimising of landscape change and fragmentation in order to facilitate likely future rehabilitation and restoration of the wetlands. This general requirement has been converted to a specific, measurable and auditable performance indicator 'the percentage of changed landscape/land cover' – that is, an outcome-based condition.

Checking and review

The checking and review elements of the HSE–MS are essential to ensuring compliance and delivering continual improvement. As well as monitoring, as discussed in the previous section, key activities include auditing, investigation of non-conformances – for example, using root cause analysis – corrective and preventative action and management review. To ensure that EIA requirements were fully integrated with these processes, the Central Environmental Compliance Database was subject to internal audits as discussed above. Non-conformances identified during the audits were then subject to investigation and corrective action in accordance with standard HSE–MS processes.

Following from the discussion in the previous section of the need to monitor the condition of the receiving environment as well as operational performance, two other initiatives were implemented to ensure that actual impacts on the environment were subject to the same follow-up processes:

- Any environmental changes or incidents – for example, fish death or weed invasion – observed by contractors in the field or by local stakeholders were recorded as incidents, using the same incident notification card as HSE incidents recorded inside the conventional borders of the project and subject to the same level of investigation and corrective action;
- Similarly, any environmental complaints received from external stakeholders and managed in accordance with external communications procedures were treated in the same way.

Both of these initiatives were examples of stakeholder engagement practices that had the potential to strengthen relationships with external stakeholders, including the local community, as well as contributing to environmental monitoring and impact management. In practice, success was somewhat limited due to factors such as lack of regulatory support, inadequate public awareness and capacity, lack of strong social infrastructure within the local communities and many other reasons. However, we believe that involving external parties in this way could be very beneficial in more conducive contexts.

7.4 Discussion, Conclusion and Recommendations

The two case studies discussed in this chapter provide some practical examples of how EIA and EMS practitioners can improve the linkages between these important and well-established, but traditionally separate tools, in the context of major oil and gas projects. It is acknowledged, however, that in both cases these initiatives were relatively small steps undertaken in less-then-ideal circumstances in which the EIA and EMS teams and activities would otherwise have been separated by a chasm, due to factors such as:

- The EIA being undertaken as a separate process (usually by external consultants) early in the project life-cycle when limited engineering information is available upon which to base the assessment of environmental risks and the identification of suitable mitigation measures;
- Related to the above, EIAs being undertaken prior to the appointment of the main project contractor, resulting in limited or no interaction between the EIA practitioners and those responsible for implementing proposed mitigations and managing environmental risks. This in turn makes it difficult for the project team to effectively apply the necessary iterative process to the development and refinements of environmental mitigation measures as the project progresses;
- A misalignment between the language of EIA and EMS, and the language of engineering design, often contributing to a failure of project teams to effectively incorporate EIA outcomes into the EMS/HSE–MS and to project design and development processes (e.g. hazard identification and risk assessment).

Two practical recommendations that emerge from the two case studies presented in this chapter are:

- EIA and EMS practitioners should become more conversant in each other's terminology and techniques, as well as the language of hazard identification and risk assessment. This will support the integration of EIA outcomes into the project HSE–MS and into the core project design and development activities. For example, it may be useful for EIA practitioners to present their findings, particularly the EMP, in the structure and language of EMS, and EMS practitioners may need to ensure that project risk assessment tools adequately reflect environmental, as well as health, safety and technical consequences;
- A Central Compliance Database should be established, which incorporates all EIA outcomes as well as other HSE compliance requirements, and which is incorporated into a broader HSE–MS Information Management System. Development of the Information Management System can commence at the early stages of the EIA study and gradually be completed as the EMS/HSE–MS is developed. It should include at least the following elements: hazards and impacts (both those identified in the EIA process and those identified during EMS/HSE–MS development); legal and other requirements (the Compliance Database); incidents and observations (e.g. any change observed in the conditions of the environment); monitoring results, non-conformities and grievances; actions and action tracking (including mitigations proposed in the EIA report as well as actions resulting from incident investigation, non-conformities, grievances etc.). Such an information system can provide a basis for tracking all potential environmental impacts in practice;
- Environmental monitoring programmes should reflect the requirements of both EIA and EMS by including both 'environmental condition indicators' as well as 'environmental performance indicators', which in turn should be divided into Operational Performance Indicators (OPI) and Management Performance Indicators (MPI). Any non-conformances identified, including environmental issues observed by external stakeholders should be subject to investigation and corrective action in accordance with the HSE–MS.

While these and other initiatives discussed in this chapter proved very effective within the limitations of this context, we believe that there are significant opportunities within the oil and gas sector to improve the robustness of project environmental management practices more substantially. We argue that this requires a fundamental paradigm shift towards a more holistic approach to environmental management throughout the project life cycle, such that EIA and its implementation and follow-up through EMS are seamlessly integrated under a broader project environmental management umbrella. In this way, EIA is never complete but rather continues throughout the project life cycle, adapting and evolving as the project definition develops.

One effect of the conventional approach to EIA and EMS is that the opportunity is missed to harness the creative capacity of project planners and designers towards improved environmental outcomes. 'Designing out' environmental risks and impacts is the most effective way to ensure compliance with EIA outcomes, and our experiences suggest that when the interaction between EIA team and project technical teams are properly established, the EIA process can add enormous value to the project design and implementation. Therefore the integration between EIA/EMS and project design and development is equally, or even more important. We offer the following inter-related recommendations to support a more holistic, integrated and effective approach to project environmental management:

- Representatives of the EIA team should be part of a multi-disciplinary technical advisory team appointed by the project owner and remaining in place from FEED to tendering and contracting with the main contractor, to the point at which project implementation is underway and the EMS is fully developed and implemented. We suggest that the following disciplines should be represented as a minimum:
  - Drilling/reservoir advisor, with knowledge of well control, drilling wastes, clustering and locating of wells;
  - Surface facility (mechanical or chemical process engineer) advisor;
  - Construction advisor;
  - Project planning/value engineering/cost engineering advisor;
  - Safety engineer and risk assessment advisor.
- The EIA team should operate as part of the project delivery team and

the terms of reference (ToRs) for EIAs should reflect this. Specifically, we suggest that services packages for 'Interactions with project design, development and implementation teams' and 'Deployment and cascading of EIA outcomes' should be added to the conventional EIA TOR and tendering documents. In practice, this means that the appointment of the EIA and other project teams should be aligned to facilitate the interaction of the EIA team with the broader project team, both to seek input into the identification of potential environmental impacts and suitable mitigation measures, and subsequently to support the design team in the development and refinement of mitigation strategies.

Building upon this point, the interactions between the EIA team and the broader project team should be planned and formalised. Appropriate mechanisms could include the following: specific EIA-focused workshops and communications meetings; EIA team participation in kick-off meetings for key contracts design workshops; and EIA team participation in design reviews. Interactions between the EIA team and the project HSE–MS team are particularly important to ensure alignment and integration. Our proposed model showing how the EIA team, HSE–MS team and design and development team is shown in Figure 7.1.

The involvement of four teams, each one with their own expertise, in the arrangement of Figure 7.1, creates four 'bridge' areas with special characteristics, as described below:

EIA–EMS/HSE–MS

- EMP
- Environmental compliance criteria
- Inputs to HSE–MS (list of requirements and procedures, etc.)
- Stakeholder engagement plan (including disclosure and reporting, compensation packages, etc.)
- Adaptive management plan
- Project closure and rehabilitation plan

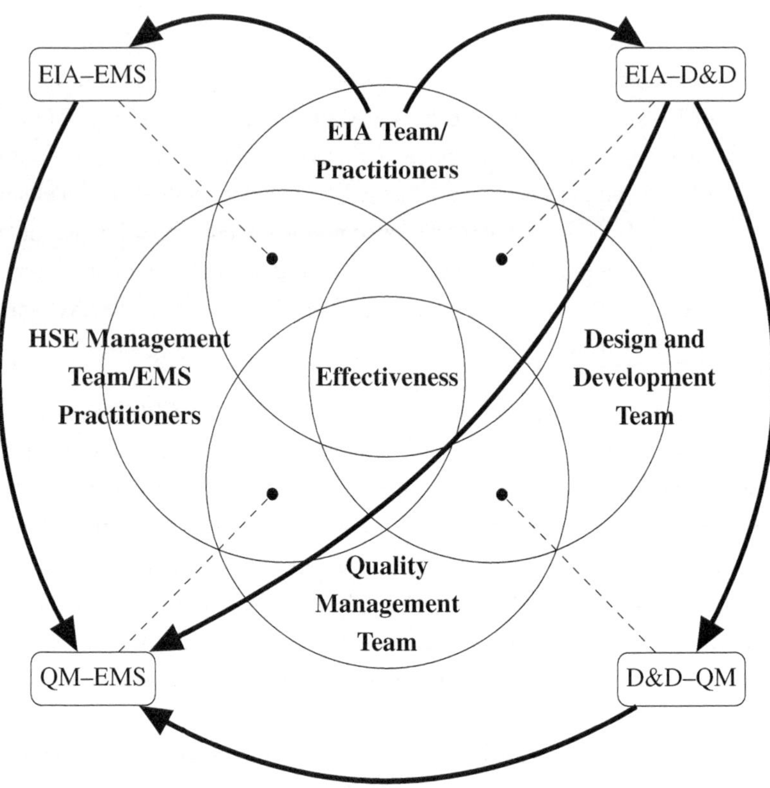

Figure 7.1 Proposed model for interactions between EIA team, design and development team(s) and HSE Management teams; the 'bridge' areas are detailed in the text

EIA–Design & Development

- Criteria for detailed site/route selection
- Design factors
- Compliance criteria
- Substantial mitigations need to be incorporated into the project design

Design & Development – Quality Management

- Quality Assurance/Quality Control plan, quality management manual and procedures
- Mechanical integrity plan (including standards for material selections, inspections, etc.)
- Management of change and process
- Contract management procedure(s)

QM–EMS/HSE–MS

- Risk assessment, HAZID, HAZOP, JHA workshops
- HSE philosophies
- Incorporation of HSE requirements into the project design
- Incorporation of HSE requirements into the construction/drilling operational manual and procedures

In closing, we hope that the practical examples offered in this chapter might provide environmental practitioners working within the oil and gas sector with some useful tools to support a gradual shift towards a more coherent and robust approach to project environmental management. Achievement of the kind of integrated, holistic model that we have proposed is a long term goal, however, and will not happen overnight. However, we hope that the ideas presented in this chapter may contribute to this journey and provide inspiration to practitioners seeking a better approach to environmental management in this sector.

8. EIA–EMS Link from the Renewables Sector

Fiona Becker[1]

This case study provides an example of how linkages can be made between an Environmental Management System (EMS) and Environmental Impact Assessments (EIA) within the Renewables Sector. The study is based on an example of how this has been adopted within ScottishPower Renewables, United Kingdom, as part of their EMS.

8.1 Setting the Scene

8.1.1 ScottishPower Renewables

ScottishPower Renewables is part of the world's largest wind power company IBERDROLA. IBERDROLA, with renewable operations in 23 countries, is the world leader in its sector by both installed capacity (nearly 13,000 MW at June 2011) and output (over 15,000 million kilowatt/hour generated over the first half of 2011) (Iberdrola, 2011).

ScottishPower Renewables became the first developer in the UK to achieve one gigawatt of electricity production capacity from wind power. The milestone was achieved after the company officially opened Arecleoch (120 MW) and Mark Hill (56 MW) windfarms in South Ayrshire in June 2011. ScottishPower Renewables, as of June 2011, has 24 onshore windfarms fully operational across the UK, consisting of more than 770 turbines. In order to achieve 1 GW of capacity, the company has invested more than £1 bn in wind energy in that last

[1]Disclaimer: The views expressed in this Case Study are those of the Author and do not represent those of the employer, ScottishPower Renewables.

decade (ScottishPower Renewables, 2011).

ScottishPower Renewables (hereafter referred to as 'the Company') are committed to minimising the impact of their activities on the environment. Although the Company had been adopting and demonstrating best practice in environmental management in regards to the development, construction and operation of their projects and assets over several years, it was recognised in 2008 by the Company that there would be benefits from the development and implementation of an EMS for their renewable power generating projects. Subsequently a Strategy was prepared and work commenced on developing an Environmental Policy and EMS. The EMS was implemented in 2009 and ISO14001 Certification for the system was received in February 2010, approximately 6 months after implementation (Certificate Reference 4005601 LRQA).

The commitment that the Company has towards sustainable development and the integration of environmental management associated with their activities has been recognised, not only by the certification of the EMS to ISO14001 but also by various external awards that include the Queen's Award, 2011. The award was presented to the Company on the basis of its commitment to sustainability, which was evidenced throughout the development, construction and ongoing operation of Whitelee Windfarm. It was also noted that the project was a best practice example of habitat management, and that the commitment to encourage local community involvement had been a strong focus.

This case study provides an example of how the Company's EMS provides linkages with the EIA process for the onshore windfarm portfolio. The onshore portfolio has been selected for this case study, as this technology, at the time of publication, is currently in all three phases of life cycle (Development, Construction and Operation) within the Company.[2]

8.1.2 Environmental Impact Assessment

EIA 'is a means of drawing together, in a systematic way, an assessment of a project's likely significant environmental effects. This helps to ensure that the importance of the predicted effects, and the scope for reducing any adverse

[2] It should also be noted that Decommissioning, the final phase in the project life cycle, is also covered by the environmental assessment process and will require similar controls to the construction phase.

effects, are properly understood by the public and the competent authority before it makes its decision' (Scottish Government, 2011, s. 8).

The EIA process can be extensive in terms of scope, resources, time and costs. Currently, certainly within the UK, this process is comprehensive and involves extensive studies, consultations and assessments. A key part of the Process is the production of the Environmental Statement (ES). The ES is the formal publication of the findings of the EIA and sets out proposed mitigation and monitoring measures for the Development under assessment.

The ES is submitted as part of the Planning Consent Process for the proposed Development. If consent is approved for the said Development, it is common practice for the associated Planning Conditions to include, where appropriate, the mitigation and monitoring measures that have been recommended within the ES. The ES is therefore a key document when considering the linkages between the EIA process and EMS.

In addition, based on project experience within the United Kingdom, it has been observed that there is an absence of a standard approach or guidance within the planning or environment profession or indeed from the determining bodies (such as the Local Authority or Government Department) or regulators (such as Environment Agency or Scottish Environment Protection Agency) to determine the effectiveness of the application of the recommended measures that have been enshrined within the Planning Conditions. It is therefore considered that there is currently a gap in this 'feedback cycle' and an opportunity for linkages to be made between the EIA process and EMSs. The following case study sets out how this is being addressed, by the developer, within the onshore windfarm sector.

8.2 Approach Adopted Through Project Life Cycle

The EMS within the Company provides the tools and mechanisms for monitoring and reviewing the consenting and legal requirements across the life cycle of a project. In addition, an integrated approach within the business is adopted in terms of project development. This involves staff from the three phases of the project life cycle – that is, development, construction and operation – to participating in the development stage. This provides the opportunity to discuss, communicate and cascade down the project environmental management requirements, such as environmental sensitivities, mitigation measures and operational restrictions to those involved in the subsequent Construction and Operational stages. This approach also provides the mechanism for a 'feedback loop' to consider what has worked on other projects for consideration in future projects. The life-cycle project stages and associated key mechanisms linking EIA and EMS are set out below, in a linear sequence of three phases.

Phase 1: Development

- Integration specialists: construction, operation, environmental
- Iterative design
- Environmental statement
- Handover
- Lessons learned
- Management reviews
- Annual compliance review
- Training and awareness

Phase 2: Construction

- Contract schedules
- Planning condition register
- Environmental site plan
- Environmental audits
- Annual compliance review
- Lessons learned
- Management reviews
- Reporting systems
- Training and awareness

Phase 3: Operational

- Site handover
- Planning condition register
- Environmental site plan
- Environmental audits
- Annual compliance review

- External annual reporting
- Lessons learned
- Management reviews
- Reporting systems
- Training and awareness

8.2.1 Development

At the Development Stage, the Company's construction, operational, technical and environmental (such as ecology) specialists are integrated into the project teams. This approach provides the specialists with the opportunity to input to the project design process, understand the site sensitivities and provide advice on mitigation measures, monitoring or habitat management plans, as required. Such measures, where appropriate, are then incorporated into the ES. These measures are generally formalised via the Planning Conditions for the development and/or other Licences/Agreements which are required to be adopted during the construction and or operational stages.

The development stage also involves a project handover phase which involves the construction team, after Consent Award, taking on the project. This phase involves the compilation of a Planning Conditions Register. The development team continue to be involved in the project post consent award, applying their project knowledge to the development prior to commencement conditions, thereby providing a seamless link between the early phases of the construction and planning conditions compliance process. All of this forms part of the environmental management process adopted by the Company.

The above integration, at the early stage of project development, is the commencement of linkages between EIA and EMS that will further develop throughout the project life cycle. This is further highlighted in Figure 8.1.

8.2.2 Construction

Due to the involvement of construction specialists at the development stage, the construction teams are already aware of the environmental management requirements for the Project when they take over the project during the handover phase. A Planning Conditions Register is compiled and is used as an internal tracking tool to record these requirements, including timescales and responsibilities.

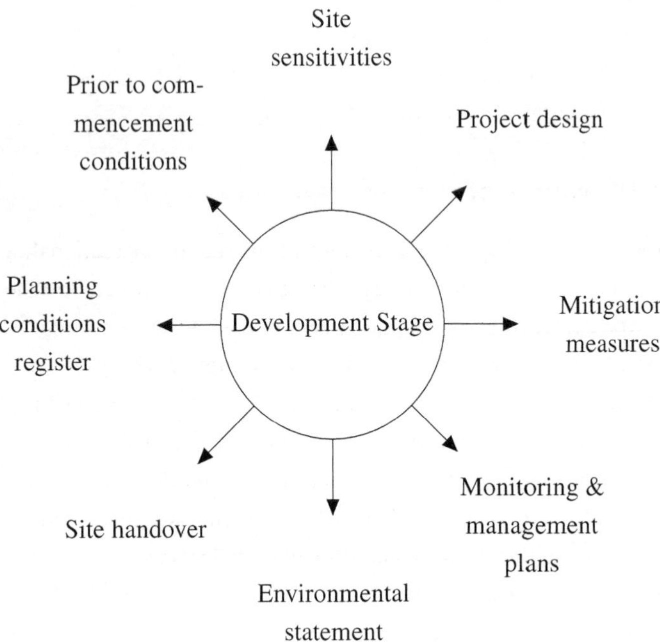

Figure 8.1 Development stage (specialist input, construction, operations, technical, environmental): EIA and EMS linkage

Based on experiences to date, in general, the majority of environmental management requirements stemming from the EIA process are required to be adopted during the construction stage. In addition, within the Company, the construction activities are contracted out to third parties such as the infrastructure contractor and turbine supplier. It is therefore imperative that such requirements are communicated and monitored during the Construction Stage to assess and monitor compliance to the planning conditions and EIA requirements. Various mechanisms have been adopted to facilitate this linkage as part of the EMS. These mechanism and linkages are set out in Figure 8.2. A key mechanism is the Site Environmental Management Plan or Construction Environmental Management Plan. These plans capture the site specific environmental requirements and are generally iterative, developing through the construction programme.

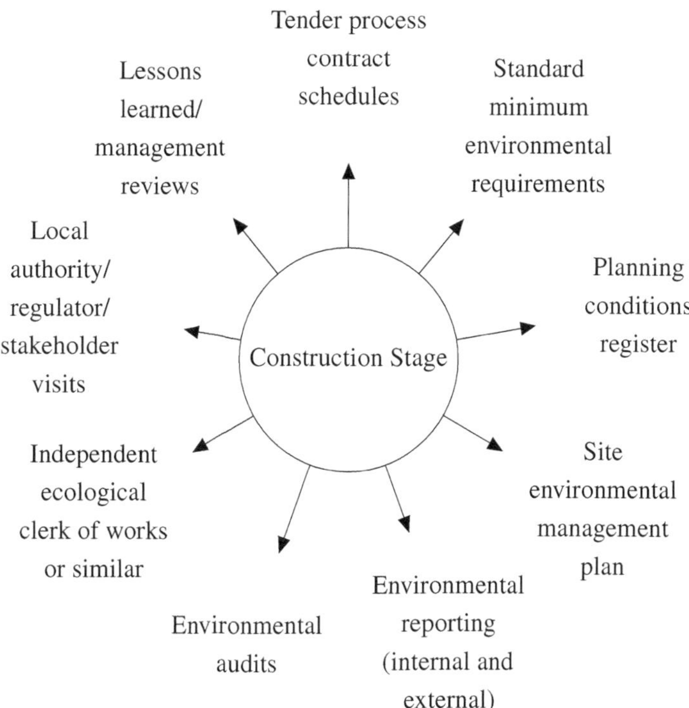

Figure 8.2 Construction: Key EIA and EMS linkages

Further details on some of the key elements of Figure 8.2 are provided in the following list.

Tender assessment and contract schedules

- Planning conditions/licences
- Environmental statement
- Environmental requirements (see below)
- Legal requirements and best practice

- Contractor assessment

Environmental requirements

- Environmental site plan
- Site inductions
- Site inspection
- Emergency response
- Procedures such as incident reporting

Environmental reporting and audit

- Environmental meetings (on site)
- Environmental incident reporting
- Environmental risks
- External reporting to stakeholders
- Internal reporting
- Environmental audits (internal and independent)
- Compliance audits

As mentioned previously, based on experience to date, the majority of planning conditions set for onshore windfarm projects relate to the construction stage of the project. This reflects the nature of the activities and the associated potential for environmental risks that require to be controlled. Within the Company, and other similar developers, the construction phase of the Windfarm project is carried out by third parties on behalf of the developer. It is therefore imperative that environmental management measures are in place to provide the mechanisms to fulfil the requirements of the EIA process (that have been enshrined within the planning conditions or associated licences) during such activities.

A key part of this process is integrating the planning condition/licence requirements within the tender and contract documentation. This is complemented by knowledge exchange on the history of the site, communication of the Company's environmental requirements via specific contracts schedules and projects being led by representative of the Company, with the project knowledge, supported by the planning and environmental specialists.

Although mechanisms are in place within the Company's system to document and communicate environmental requirements, these must be checked. This is carried out via various means, some of which are highlighted below.

- Independent audit/inspections – on site, daily/weekly/monthly via ecological clerk of works or similar
- Stakeholder inspections/visits – from the local authority, environmental regulator
- Integrated project team – site project team and associated reporting/communication process
- Internal audit – periodic visits, compliance review

8.2.3 Operational

During the operational stage of an onshore windfarm the scope of site activities is more limited than during construction, however there are still associated potential environmental risks that must be controlled. Along with the construction stage, operational impacts and associated mitigation measures or controls are included within the ES. Linkages back to the EIA stage are therefore also relevant during the operational stage. This is adopted within the Company as an inherent part of the ongoing site management controls such as work control procedures and site specific requirements. Tools used to facilitate this are shown in Figure 8.3.

8.2.4 Example of Linkage: Habitat Management

Habitat management is a key area that the Company are committed to developing. At the end of 2010, the Company had approximately 80 km^2 of habitat management areas. These areas covered 11 windfarms with a variety of objectives including the provision of habitat improvement areas for black grouse, wading birds and golden eagle.

Progress on meeting the objectives of the habitat management plans are reviewed and recorded internally on an ongoing basis. In addition, where required, progress is also reported externally via Habitat Management Groups. External representatives on these groups may include those from Scottish Natural Heritage, The Royal Society for the Protection of Birds (RSPB) and the Forestry Commission. The groups offer the facility to review how objectives are being met, discuss how effective the measures are proving to be and revise these measures if necessary. The opportunity for this ‘knowledge exchange’ also provides a mechanism for group members to consider the findings in relation

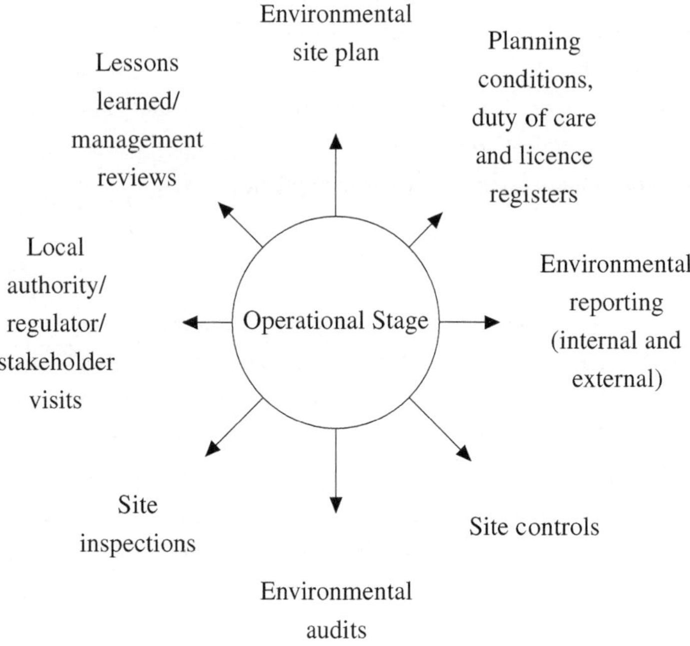

Figure 8.3 Operation: detail on key linkages

to the development of other relevant habitat management plans. One of the groups, Blacklaw Habitat Management Group, has a dedicated extranet web site specifically for members.

8.3 EM Tools – Examples

Section 8.2 highlighted key examples of linkages between EIA and EMS. However there are additional tools that support these linkages, examples of which are provided below.

8.3.1 Environmental Training and Awareness

A key part of successful environmental management within business is ongoing environmental training and awareness of staff. It is considered that this has been a key factor within the Company in regards to the way that environmental matters are embraced Company wide. This is facilitated by various means including the following:

- Environmental awareness training (ranging from two-day sessions to one-hour Tool Box Talks)
- Environmental news bulletins
- Environmental legislation updates
- Communications sessions
- Promotional weeks (e.g., Green Week)
- Central reporting systems for environmental data

The above, where appropriate, often have linkages between the early development stage and associated EIA process, with environmental management at the site level such as legal compliance, audit findings, management controls, success stories (such as habitat management objectives, awards) and data management.

8.3.2 Project Reporting Protocol

A key question that is often raised to link the EIA process with what actually happens in practice is in regard to the effectiveness of the on-site mitigation measures. In addition to the points raised in Section 8.2, one way that this can be assessed during and after the construction and operational phases of development is by reviewing whether any major environmental incidents have been reported and whether any regulatory notices been served or enforcement actions taken by the regulator or planning authority. Such occurrences would highlight instances where the required control measures have not been put in place, have not been applied or are not working effectively or where improvements/alternatives are required.

To enable the above to be carried out a project reporting protocol is required. In addition, the review of reported environmental incidents on a regular basis also gives an indication of the effectiveness of the site control measures on site and linkages back to the EIA process. This can also be fed back via lessons learned sessions to appropriate project staff. Such a mechanism is in place within the Company.

8.4 Conclusion

The renewables sector is constantly expanding and government targets are becoming increasingly ambitious. The 2009 Renewable Energy Directive sets a target for the UK to achieve 15 per cent of its energy consumption from renewable sources by 2020. The Scottish Government have set a national target for 80 per cent of Scottish electricity consumption to come from renewables by 2020 – recognising the extent of Scotland's important renewables resource.

Although the sector, in comparison to some other industries (such as mining, chemicals and traditional power generation), have fewer processes to control, environmental management is still a key consideration. In particular, as development in offshore wind, wave and tidal mature, the nature for potential environmental impacts will diversify and there will be increasing requirements for environmental management controls.

It is considered that the adoption of an EMS, with inherent linkages back to the development stage, will facilitate and support the development of such facilities in an environmentally acceptable and sustainable manner. Such a system should be designed to include the mechanism for a 'feedback loop' to the planning conditions and associated ES requirements. This would then provide a mechanism for the developer to check they are complying with the project requirements throughout the project life cycle.

However, in order to complete the 'feedback loop', a mechanism requires to be put in place within the determining body or regulator to support the process. It is considered that an external mechanism is required so that knowledge is being cascaded via two routes, not only internally within the developer's company, but also externally within the determining bodies and regulators, so they are also benefiting from the 'lessons learned' and 'knowledge exchange process', in addition to providing a mechanism to audit compliance against the various planning conditions and ES requirements. It is considered that a combination of the two routes would provide an overarching mechanism to link EMS and EIA. This could be facilitated regardless of the industry or geographic location.

9. EIA–EMS Link from the Waste Management Sector

Lisa Palframan

Alongside the increasing recognition of the need for resource efficiency, the delivery of infrastructure to manage waste with minimal environmental impact is recognised as a major priority (DEFRA, 2007a; UNECA, 2009). Public concern over the possible health and social impacts of waste management sites has also become apparent. Environmental tools are therefore needed to address both environmental and social concerns about waste management. Environmental Impact Assessment (EIA) and Environmental Management Systems (EMS) are capable of collectively addressing these concerns, to enable infrastructure to be delivered in a way which minimises impact. This chapter outlines the environmental impacts of waste, the types of waste management infrastructure being delivered within the European policy context and the planning and consenting of waste projects in Europe. It then explores how EIA and EMS can be used together in the case of a multi-site company operating a landfill site in the UK.

9.1 The Environmental Impacts of Waste Management

There is a strong link between economic growth and waste production. As societies prosper, consumption levels rise and more waste is generated. In many countries, there is increasing awareness of the need to minimise waste produced and increase recycling levels. Indeed within Europe, comprehensive legislation has forced Member States to ensure businesses and consumers reduce the amount of waste produced. The End of Life Vehicles Directive (2000/53/EC) and the Packaging Waste Directive (94/62/EC) are two examples which aim to reduce the environmental impact of waste and identify responsibilities for recycling and sound management of waste. Meanwhile, the Landfill Directive (1999/31/EC) has placed various restrictions on the volumes and types of materials sent to landfill, for example banning the disposal of tyres and gradually phasing out the disposal of biodegradable waste.

The EU Waste Framework Directive presents the Waste Hierarchy as a guide to the environmental impact and therefore desirability of waste management options (European Commission, 2008). Waste prevention should be prioritised but if this is not possible, materials should be prepared for re-use. Where this cannot be achieved, materials should be recycled, or less ideally, the energy recovered. Final disposal without recovery is the least preferred option. With increasing environmental impact down the hierarchy also comes the need for physical infrastructure to process the waste and derive value from it where possible. Mass recycling of unsorted recyclables from businesses or households may require the construction of a Materials Recovery Facility (MRF) to enable the efficient mechanical sorting of materials into separate streams. Energy recovery may be achieved *inter alia* through incineration with a district heating scheme and anaerobic digestion with biogas collection. Without energy recovery, landfill and incineration are considered as the least desirable final disposal. However, Government policy recognises a role for properly managed landfill for dealing with the wastes from which no further economic value can be derived (DEFRA, 2007a).

Depending on the national legal framework, the proposed development of new waste management facilities may require that an environmental impact assessment be carried out. For example, the European Union EIA Directive (85/337/EEC as amended) requires an EIA for certain hazardous waste facilities and certain non-hazardous facilities exceeding 100 tonnes per day throughput as listed on Annex 1. Under Annex 2, it requires EIA for some other waste management facilities where the Member State establishes that the project is likely to have significant effects on the environment (see Table 9.1).

Table 9.1 Summary of EIA Directive annexes as relevant to waste projects

Annex	Requirement for EIA	Projects listed
1	EIA needed for every project	Waste disposal installations for the incineration, chemical treatment or landfill of hazardous waste. Waste disposal installations for the incineration or chemical treatment of non-hazardous waste with a capacity exceeding 100 tonnes per day
2	EIA needed where the Member State establishes (on a case by case basis or through the use of thresholds) that the project is likely to have significant environmental effects	Waste disposal installations for the incineration or chemical treatment of non-hazardous waste with a capacity exceeding 100 tonnes per day

9.2 The UK Context for Waste Management and the Role of Environmental Tools

In the past, the UK has lagged behind other European countries in the recycling and sustainable management of its waste. In recent years however, in spite of an increasing population and rising number of households, it has managed to stabilise the amount of local authority-collected waste (including household waste) being produced at around 34.3 million tonnes by 2008 (DEFRA, 2011a). Of this, nearly 40 per cent is now recycled or composted, representing significant progress towards the EU Waste Framework Directive 2020 target for 50 per cent of waste to be recycled, composted or re-used. In the industrial and commercial sector, the volume of commercial and industrial waste produced was reduced by 29 per cent from 2002/3 to around 48.0 million tonnes by 2010 (DEFRA, 2010). Contributing to this change however was the decline of resource-intensive manufacturing and heavy industry particularly in the north of England.

The UK has traditionally relied on landfill as the main means of waste disposal, although incineration has played an important role in some parts of the country. The decline in available space for landfill especially near urban areas and the introduction of the Landfill Directive led to the implementation of alternative approaches to waste management (DETR, 2000). New European legislation has also influenced the construction of facilities using alternative technologies. Tighter controls on emissions during the late 1980s led to many existing waste incinerators being closed (Petts, 1994). More recent legislation has required that energy recovery is incorporated into new incineration schemes and that more challenging pollution prevention requirements are met. For example, the Waste Incineration Directive 2000/76/EC set stringent limits on emissions of substances such as heavy metals and nitrogen oxides (NO and NO_2) for both new and existing plant. This has necessitated the upgrading of existing facilities in addition to the construction of new facilities that meet the increasing demand. New schemes have been brought forward, with nine of the nineteen UK incinerators operating in 2007 having been established during the preceding seven years (DEFRA, 2007b). While the amount of waste going to landfill has nearly halved over the past ten years, 2007 figures show that landfill continues to be the final destination for around 56 per cent of the waste produced (DEFRA, 2011b).

The UK Government is dependent on the private sector bringing forward new waste management infrastructure and while historically levels of private sector investment have been relatively low, there is now recognition of the need to intensify the rate of this investment (DEFRA, 2007a). In commercial terms, the UK waste sector is estimated to include around 1000 companies (Grant Thornton, 2008) but it is dominated by twenty large players, which account for combined revenues of around £5 billion (Wilson and Pearce, 2009). These companies usually operate multiple landfill and energy from waste sites. Such sites are subject to regulatory scrutiny for both planning and pollution control purposes due to their potentially significant environmental impacts

Planning applications for relevant projects must meet the requirements of the EU EIA Directive, which has been transposed in the UK using a range of Statutory Instruments. Of particular relevance to waste projects are the Town and Country Planning (Environmental Impact Assessment) Regulations 2011 which apply to projects requiring planning permission. Similar regulations apply in Wales, Scotland and Northern Ireland. These regulations require that when submitting an application for an EIA Development (as defined by Schedules 1 and 2 of the regulations which follow Annexes 1 and 2 of the Directive), proponents must also submit an Environmental Statement which reports on the EIA. This must include descriptions of the development, alternatives studied by the developer, the environmental baseline, the likely environmental effects and proposed mitigation measures. A non-technical summary is prepared to facilitate public participation.

In addition to obtaining planning permission, operators of certain industrial facilities in England must also apply to the Environment Agency (EA) for an Environmental Permit (EP), under the EU Integrated Pollution Prevention and Control (IPPC) Directive (2008/1/EC). A similar regime operates in the other UK countries. The EP specifies activities permitted under the licence and environmental outcomes that operators must achieve, such as emission limits and appropriate waste storage. Sites may usually select the technique or approach that is considered most appropriate to achieve compliance. Standard permits are available to operators who are able to comply with fixed rules for common activities, otherwise an application for a bespoke permit must be made.

National planning guidance encourages parallel applications for planning consent and EPs, explaining that 'the controls under the planning and pollution control regimes should complement rather than duplicate each other' (Communities and Local Government, 2004). In theory, the planning consenting decision should be concerned with IF a project should go ahead and the pollution permitting system should concentrate on HOW that development should proceed, in a way that prevents and minimises the release of substances to the environment (DEFRA, 2007c). Ideally this should represent a single decision because the two elements are interdependent. This would enable the planning (including EIA) and permitting system to be used jointly to achieve effective environmental protection.

Researchers investigating the planning and pollution control systems in Scotland have noted inefficiencies in this relationship (Environmental Resources Management, 2004). The potential role of EIA in enabling effective communication between planners and the environmental regulator is not always taken up, suggesting the need for improved guidance. A study of the costs and benefits of the EIA Directive at a European level recognised overlaps with the IPPC Directive but stated that they had been difficult to rectify because of the need for two consents (GHK, 2008). It was suggested that EU Member States had found difficulties in linking EIA and IPPC, although Germany had been able to place both processes under the same regulation and harmonised the project thresholds that determine permitting requirements.

In England, all operators applying for an EP must show that they will implement a formal EMS (Environment Agency, 2011a). An EMS can be described as '... that part of an organisation's management system used to develop and implement the environmental policy and manage its environmental aspects' (ISO, 2004). It ensures that the main environmental impacts of an organisation are identified, managed and reviewed, which should lead to continual improvement in environmental performance over time. Operators of larger and more complex activities need to operate an externally certified EMS and must provide a summary of their management systems arrangements when making their permit application (Environment Agency, 2011b). Indeed, the lower risk presented by sites which operate an externally certified EMS is recognised by the Environment Agency in their OPRA (Operational Risk Appraisal) scheme. This places installations into bands and offers a reduced inspection frequency and lower fees to sites with high scores against five attributes. One of these attributes is Operator Performance which assesses management systems arrangements for operator and maintenance, competence and training, emergency planning and organisation (Environment Agency, 2011c). An externally audited EMS directly results in a higher score for 'organisation' and the rigorous management arrangements required by an externally audited EMS for the other areas also contributes to a better score.

EMS could potentially play an important role in ensuring EIA outputs are addressed. EIA findings might contribute to the register of aspects and impacts or inform a risk-based internal audit programme. Mitigation measures could be addressed through the programme of objectives and targets or through the monitoring system, translating generic recommendations into implementable procedures or instructions and giving reassurance to stakeholders that the identified impacts will be effectively managed (Sánchez and Hacking, 2002). Likewise, EMS for an existing site can generate environmental information that could be useful when carrying out EIA for a similar new site (Glasson et al., 2005). Marshall (2004) showed that a single company overseeing development from planning to operation made effective use of a company-wide approach to EIA and EMS. However, researchers investigating the link between EIA and EMS have noted a number of barriers to linking the tools more closely, relating to the legal/policy framework, the project process, practitioner issues and stakeholder attitudes (Palframan, 2010). Slinn et al. (2007) in particular noted the challenges in the case of companies planning business park developments which were then developed, owned and/or occupied by different organisations.

The interaction of the EIA, EMS and the EP in the management of the impacts arising from a waste facility will now be discussed by exploring a case study of the Pilsworth South Landfill Site. The case study highlights how the EIA carried out for the extension of this site has made use of information from past experience as captured by the EMS and informed the application for the variation of the permit. The use of a management plan prepared for the EP application has enabled a strong link to be made between EIA and EMS. The applicability of similar arrangements for other companies and sectors will be considered.

9.3 Case Study: Pilsworth South Landfill Site

Viridor operates 324 recycling and waste management sites in the UK including 22 operational and 25 closed landfills (Viridor, 2011) and is the fifth largest UK waste company by revenue (Wilson and Pearce, 2009). It was the first UK waste company to gain an ISO 14001 EMS certificate covering all its major sites, including Pilsworth (Viridor, 2011). The EMS is part of a corporate Business Management System which also meets ISO 9001 as a quality management system (QMS) and BS OHSAS 18001 as a health and safety management system. Viridor takes a common approach to meeting many of the requirements of these certifications, for example in publishing a Corporate Responsibility Policy, having a common process for aspects identification/risk management and a joint Management Review process. A central improvement programme is drawn up which includes corporate environmental objectives and targets. All sites are expected to make progress on these, although they are also encouraged to set their own targets to ensure that actions taken are locally appropriate.

In December 2010, Viridor submitted a planning application to the Bury Metropolitan Borough Council to extend an existing landfill site known as Pilsworth South and located in Greater Manchester, UK. An extant planning permission allows the quarrying of sand and gravel with landfilling of waste being used as part of the site restoration since 2006. Under the new plans, vertical and lateral extensions would take place within the existing permitted site boundary and increase the capacity of the landfill site by an additional 1.8

million m^3 of waste. This would allow the landfill's lifespan to be extended by up to five years with the site being gradually restored as a country park and recreation area. Although the landfill in its current configuration is not due to be completed until 2023, feasibility studies found that engineering changes were needed imminently in order to facilitate the void increases.

The application was accompanied by an Environmental Statement (ES) needed to fulfil the requirements of the EIA Regulations as a waste disposal project (Schedule 2) likely to have significant environmental effects. The main potential impacts were agreed with the council during scoping and detailed in the ES. These included ecology/nature conservation, noise, odour, dust, drainage, geology/ground conditions and landscape (Wardell Armstrong, 2010). The Council specifically requested in their scoping response that any information gathered as part of the monitoring process for the previous consents should be integrated within the studies.

The ES was prepared by environmental consultants who were appointed by the company's engineers. These same consultants had produced the studies to support planning applications made previously for the site in 2003 and 2006. They worked closely with site staff to ensure that the ES reflected actual procedures employed and the mitigation measures already in place. They also established that any further measures recommended by the consultants could be delivered. Consultation with residents and other interested parties was carried out at an early stage, including through the site's existing Pilsworth South Liaison Group and a public exhibition. Overall, the EIA concluded that there were no adverse environmental impacts of long-term or over-riding significance and stated that benefits were likely arising from the plans for restoration. At the time of writing, the application is awaiting determination.

The corporate approach to the environmental management of facilities subject to EIA is set out in Figure 9.1 and will now be discussed specifically with regard to extensions to existing development, as in the case of Pilsworth South. Key to the delivery of EIA mitigation measures and commitments in each case of EIA development by Viridor is the Operations and Development Management Plan (ODMP). This is prepared for each site in advance of an application for a new EP or variation to an existing one. It acts as an interface between the site-specific EIA (where one is required) and the corporate BMS. Operators applying for a bespoke permit must provide an environmental risk assessment and set out how they will comply with the general and technical Environment Agency standards applicable to their operation. The submission of the ODMP enables this requirement to be met. It builds on the company's experience from similar sites (and the actual site in the case of extensions to existing sites) and includes an explanation of management arrangements, together with methods and strategies to deal with significant issues such as waste acceptance, leachate, surface water, groundwater, landfill gas, litter and noise. It is prepared iteratively with the EIA and submitted to the EA at or shortly after the planning application is made to the local authority. The EA uses this information to develop permit conditions.

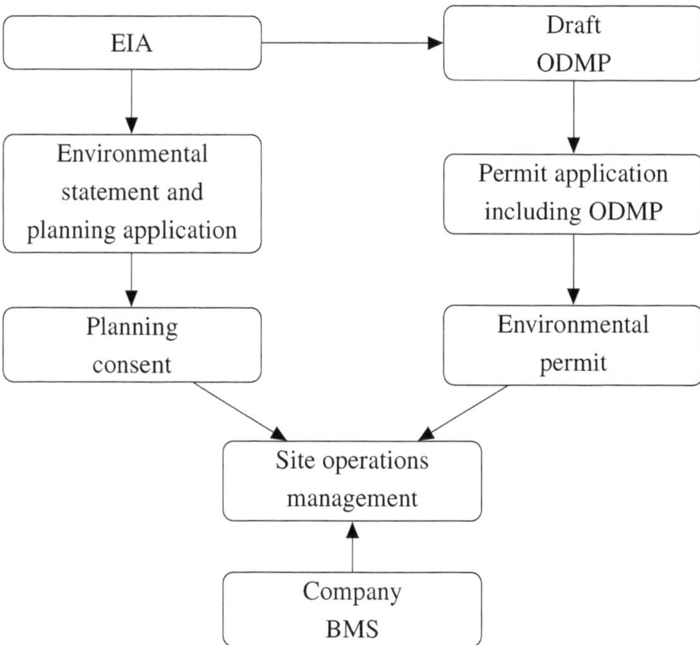

Figure 9.1 The interaction of the EIA, environmental permit and business management system (BMS) in the management of Viridor's operational facilities

In the case of Pilsworth South, the ODMP is currently being prepared. Should the planning application be consented and an EP granted, the management of environmental issues at the site will be carried out in accordance with both the site-specific ODMP, which fully reflects the EIA findings, and the corporate BMS. The corporate BMS requires that working procedures are drawn up for each site and these reflect the ODMP arrangements. These procedures give day-to-day guidance to site staff based on activity rather than environmental issue and provide a basis for local BMS targets if desired. Incorporating procedures into an ISO 14001-certified BMS keeps them 'live'. It provides assurance that responsible staff have been trained to meet the requirements laid out, for both routine operations and emergencies. Compliance with procedures is audited and the procedures themselves reviewed at intervals to ensure they remain appropriate.

Internal checking of these arrangements is twofold; auditing of site practices for compliance with the company BMS, which is carried out by a central company audit team, and auditing of compliance with the EP, EIA and working procedures, undertaken by site environmental staff and central Compliance Managers. This means that auditing is undertaken against both company-wide and site-specific requirements, providing robust risk management arrangements. The sites regularly report their environmental performance to the EA in accordance with their permit conditions and submit an annual environmental report. At a corporate level, an Executive Committee reviews progress of the company against the corporate objectives every four months. An externally verified Corporate Responsibility report is published annually (Viridor, 2011).

Information collected within the BMS helped inform the EIA studies. For example, the company BMS incorporates an Incident Database which helps fulfil the Records (section 4.5.4) and Management Review (section 4.3) requirements of ISO 14001. Records from this database enabled the ES to confirm that no complaints about dust had been received since waste disposal operations commenced on the Pilsworth South site (Wardell Armstrong, 2010). Actions to minimise dust had been agreed for the previous application as part of the EP and the ODMP details the procedures to be followed for reporting of dust monitoring records and in the event of any complaints. Monitoring of selected ground gases had been a requirement of the EP from 2006. The results were presented in the ES and used to assess the risk of ground contamination, making use of monitoring data as requested by the planning authority in their scoping response.

9.4 Analysis and Implications

Major waste companies may need to routinely undertake EIA due to the potential of facilities such as landfill and incinerators to cause significant environmental impacts. EPs are also required to operate such sites so operators may find, as in England, that the regulator considers a formal EMS essential to control these impacts. In contrast to many other sectors such as retail and housing, the nature of their activities means companies in the waste sector are likely to plan developments or re-developments which they then construct, operate and de-commission (e.g. site restoration for landfill), giving them control over the whole project life cycle. In these cases, it makes sense to have a corporate approach which links EIA for new developments with the EMS, regardless of whether a corporate EMS is operated or whether each site has its own system.

For Viridor, the fulfilment of the environmental requirements of the consenting authorities is indeed a routine issue within the business, involving environmentally focused staff both on-site and within the central company team working together. Under these circumstances, the full integration of environmental considerations into the planning and management of developments is highly efficient. The ODMP is prepared for the EP application but acts as an important interface between the EIA and the BMS, creating a more detailed strategy from the general EIA mitigation and monitoring measures to help inform the regulator's decision on the granting of the EP. This strategy can then be 'translated' into more practical instructions in the form of BMS procedures for addressing key impacts (Sánchez and Hacking, 2002). These guide site staff in carrying out day-to-day tasks.

The ODMP could be described as an Environmental Management Plan, a document which is sometimes considered as a less formal substitute for EMS, which helps ensure that proposed mitigation measures are implemented. Marshall (2004) describes environmental management plans as an 'EMS-lite' approach, because they are used by some companies as a simpler and less bureaucratic alternative to a formal EMS. In Viridor's case, the ODMP links to the externally audited EMS rather than substituting for it, providing a more robust approach.

Many of the barriers to linking EIA and EMS identified from previous work no longer apply in the case of Viridor (see Table 9.2), due to the characteristics of the consenting arrangements and the involvement of the company team through the project planning and development. While the EIA development discussed is an extension to an existing project, the corporate BMS means these arrangements could easily be adapted to similar new projects.

Table 9.2 Perceived barriers to linking EIA and EMS – adapted from Palframan (2010)

Type of Barrier	Example	Reference	Relevant to Pilsworth
Legal and policy framework	Voluntary basis of EMS providing little incentive for uptake	(Slinn et al., 2007)	EMS is not voluntary because the environmental regulator requires a formal system is in place and there is a strong expectation that EMS operated by major sites will be externally certified
Process/technical issues	EMS orientated towards day-to-day activities, environmental implications of new development not considered	(Marshall, 2004)	The ODMP provides input from environmental management of new development towards the day-to-day management of ongoing activities; the BMS helps manage compliance with planning requirements during the construction phase
Practitioner issues	Different personnel undertaking EIA and EMS for any given project	(Ridgway, 2005; Sánchez and Hacking, 2002)	While the ES is prepared by consultants, they work with site environmental staff and the BMS co-ordinator as the ODMP is developed iteratively with the EIA, ensuring continuity
Proponent and stakeholder attitudes	Reluctance of proponent to put resources into operational management before the outcome of the application is known	(Slinn et al., 2007; DEFRA, 2007c)	The requirement to provide details of proposed management arrangements within the permit application incentivises the investment of effort in preparing operational management arrangements whilst the EIA is still being prepared

Joining up the EIA and EMS may give greater confidence to the company and the regulators (both the planning authority and pollution regulator) that issues identified during EIA are fully addressed during on-site operations. This should help improve management of environmental risk and enable the consenting process to proceed more smoothly, giving greater certainty to the developer. It should make more efficient use of existing knowledge and information compared with approaches which treat EIA and EMS separately. It could also facilitate the sharing of information across the company, for example in identifying common problems and solutions. The EP application arrangements place the ownership of EIA mitigation measures firmly with the developer, which should enable more efficient and effective delivery.

9.5 Moving Practice Forwards

As nations try to improve resource efficiency, it is also vital that residual waste is managed in an environmentally sound manner. Within Europe, increasingly stringent environmental protection requirements have necessitated new and upgraded landfill facilities. These often undergo EIA to assess the likely environmental effects, identify mitigation measures and provide an opportunity for public participation as part of the spatial planning process. In England, pollution control requirements mean major operational waste sites must implement an EMS and this is secured within the permit conditions. An EMS enables the environmental impacts predicted during EIA to be actively managed during site operations.

Previous studies have suggested that despite the environmental protection role of EIA and EMS, limited connection is made between these tools in practice. However, the case of Viridor as operator of the Pilsworth South Landfill Site shows that it is possible and indeed preferable for developers to take a unified approach. An operations management plan is prepared iteratively with the EIA. This details how the operational impacts of the development will be managed within an EMS framework to meet the anticipated permit conditions. The corporate BMS sets a company-wide framework for regulatory compliance and continual improvement in environmental performance, providing an additional level of assurance to stakeholders.

Despite criticisms of the separation between spatial planning and pollution control in many European countries, it can be argued that this approach allows the consenting body with the most appropriate expertise to thoroughly examine a proposal to establish if (planning) and how (pollution control) it should proceed. Major companies which plan and operate the types of facilities which are subjected to both processes can optimise their resources by preparing their applications iteratively and making use of EMS to deliver on their EIA commitments. Since public scrutiny of waste and other highly visible facilities is unlikely to diminish, the business case for joining up EIA and EMS will continue to strengthen.

10. EIA–EMS Link from the Flood Risk Management Sector

Karl Fuller, Claire Vetori, Bruce Munro, and Kevin House

The Environment Agency for England and Wales includes, amongst its functions, oversight of flood and coastal erosion risk management (FCRM). Direct responsibility is held for flood risk management for main rivers (larger watercourses) and the coast. This involves the planning, construction and maintenance of flood risk management projects.

This chapter sets out the background against which the Environment Agency undertakes flood risk management activities. This includes environmental duties and obligations set out in legislation and government guidance, and environmental sensitivities that are inherent in the fluvial and coastal environment of England and Wales. We then explain the framework under which environmental assessment and environmental management is implemented, before describing in more detail how this operates in practice in the implementation of a project. The chapter concludes by discussing how the Environment Agency passes lessons learned from one project to other similar projects and identifies areas in which the environmental assessment and management framework could be further strengthened.

10.1 Background

The Environment Agency was established in 1995 by the Environment Act.[1] The Environment Agency's principal aim and objectives in discharging its functions are set out in Regulation 4 of the Act; to protect or enhance the environment, and to contribute towards the objective of sustainable development.

The Environment Agency acts in a variety of functional roles in delivering its duties, acting as an operating authority, a regulatory authority and a licensing body. As an operating authority in relation to flood and coastal erosion risk management, the Environment Agency appraises, designs, builds and maintains new capital schemes to manage flood risk for people, property and critical infrastructure.

Flood risk management projects can provide a certain level of protection from flood damage for important environmental features, for example valuable heritage assets. However, they can also create conflicts with the need to protect and conserve the natural environment.

This fine balance of protecting people, property and the environment is achieved by the management of environmental risk throughout a flood defence scheme's life cycle; from the identification of the need to take action to manage flood risk, selection of the most appropriate flood risk management option, through developing a design to the construction and operation of a project. The Environment Agency also seeks to deliver environmental improvements or enhancements to an area, wherever it is practical to do so. This may include, for example, creation of new wetland features, or improvements in local footpaths and access, as part of our schemes.

Creating new flood defence infrastructure in fluvial, estuarine or coastal locations can present a unique set of circumstances, in terms of the range of technical, socio-economic and environmental constraints that need to be overcome. For example, the creation of engineered coastal defences sufficient to protect coastal communities can potentially result in adverse impacts on sensitive intertidal habitats such as saltmarsh. The sensitivity of coastal habitats in England and Wales is recognised by European Community (EC) legislation,

[1]Environment Act 1995 c.25.

including the EC Habitats.[2] and Birds Directives[3] Since many of these areas lie adjacent to vulnerable coastal communities, new defences often need to be located within or immediately adjacent to these protected habitats, meaning potentially damaging effects can be difficult to avoid altogether.

Such schemes can also generate a wide range of potential environmental impacts, such as on sensitive archaeology, valued and protected landscape features, noise impacts and disruption to the local community and businesses during construction. Recording these impacts, establishing their significance, developing mitigation to reduce or avoid impacts, and ultimately ensure legal compliance is the role of the Environmental Impact Assessment (EIA) process. As part of our statutory obligations 'to protect or enhance the environment', we also strive through EIA to go beyond protection and deliver environmental improvements wherever possible. The EIA process is therefore not seen as a standalone exercise, but as a tool to inform and influence the design process so that, ultimately, more sympathetic flood defence schemes that create a better place for people, communities and the environment can be delivered.

10.2 Risk Management and the Environment

The environmental factors that need to be taken into account in the development of a flood risk management solution are characterised as environmental risks. These may relate to risks to the delivery of the projects, risks to sensitive environmental receptors, risks of non-compliance with environmental legislation or risks to the continued provision of services provided by ecosystems. The benefit of defining these in the language of project management is that they become an integral part of the delivery of the project, rather than an environmental add on.

The approach to managing risk often helps us focus on options for managing environmental risk and delivering better flood risk management. EIA plays a key role in identifying and quantifying risks. It leads to the adoption of measures for managing those risks, thereby enabling the successful delivery of environmental outcomes.

[2]Council Directive 92/43/EEC on the Conservation of natural habitats and of wild fauna and flora (as amended).

[3]Council Directive 79/409/EEC on the conservation of wild birds (as amended).

10.2.1 EIA and the Appraisal of FCRM Schemes

EIA may be undertaken as a regulatory requirement, typically under land use planning or land drainage legislation. However, even where this is not the case a more bespoke process is applied to manage any key environmental risks. This may involve a less detailed assessment than a formal EIA (e.g. based on published information only) or focus on how the key risks will be managed to ensure effects are not significant, rather than undertaking a formal assessment. Where a more bespoke assessment process is followed, we refer to the document produced as an 'environmental report' to draw a distinction from an Environmental Statement required by regulation.

In all cases EIA is part of a formal appraisal process for which the framework is provided in guidance published by the Environment Agency. The Department for the Environment, Food and Rural Affairs sets out the governments policy on the appraisal of flood risk management including the requirement to consider the environmental, and other, impacts of all proposals: 'Environmental appraisal techniques, using the structured methodology employed in a Strategic Environmental Assessment or Environmental Impact Assessment should be utilised to describe the full range of impacts on the human, cultural, historic and natural environment of all options.'[4]

The appraisal process, including environmental assessment, is the first step in avoiding or minimising the adverse effects of flood risk management. After identifying the problem, the process requires the identification of alternative solutions that are evaluated for their effectiveness in reducing flood risk, together with their environmental, social and economic performance. Government policy and the National Flood and Coastal Erosion Risk Management Strategy[5] promote the development of solutions that work with natural processes on the basis that they are more sustainable: as well as providing better environmental outcomes, they rely less on continued maintenance and do not create the expectation of future expenditure when the infrastructure needs to be replaced. Flood risk management does not therefore always result in development. For example, withdrawing maintenance on river banks where there are no properties

[4]HM Government (2009), Appraisal of flood and coastal erosion risk management: A DEFRA policy statement, Department for Environment, Food and Rural Affairs, London.

[5]Environment Agency (2011), *A National Flood and Coastal Erosion Risk Management Strategy for England*.

at risk can create washlands that will contribute to reducing flood risk in more populated areas downstream.

However, in many cases there are few opportunities to adopt natural solutions that will reduce flood risk to an acceptable level. There is also uncertainty regarding the performance and reliability of some natural solutions and even where they are feasible, they may only be delivered through a planned intervention. For example, the establishment of intertidal habitat to provide a natural coastal defence may require the construction of new realigned defences inland and an engineered breach of the existing defences. Consequently, most flood risk management, where we are able to take action, involves engineered solutions entirely or in part.

The appraisal of options requires the gathering of a considerable amount of information on their relative performance, the associated economic benefits, costs of delivery and the associated environmental effects. In some cases environmental factors are monetised to be incorporated into the economic appraisal. As a result, by the time a preferred option has been identified, the EIA is already significantly advanced and there is a good understanding of the environmental risks with taking it forward. Similarly the mitigation and management measures that are likely to be required are under consideration and form part of the business case to demonstrate that the solution is likely to be acceptable and can be delivered. In most cases, contractors are involved in the appraisal process and particularly with the development of the preferred solution. This helps establish that the mitigation and management measures can be implemented and also provides an indication of the associated costs. We refer to this as early contractor involvement.

Business cases are developed on the basis of an outline design and are subject to independent scrutiny by a quality assurance board before they are recommended for approval. Detailed design follows with continued contractor involvement.

10.2.2 Environmental Impacts of FCRM Schemes

The impacts of flood risk management projects, like any other, depend on the nature of the project and the sensitivities in any particular location. Nevertheless, there are some factors that are common to most schemes.

The majority of impacts associated with FCRM are related to the construction phase. Noise and transport impacts and impacts on buried archaeology are typical examples. Other impacts can be integral to the existence of defences. For example, they can have a significant impact on the character of the landscape, particularly in urban areas. While the structure may function as a flood defence on an occasional basis, sometimes only once in decades, the remainder of the time it is part of the fabric of the urban space and can significantly contribute to, or detract from, the character of the location. Other long-term impacts are less easily perceived, but can be significant over the lifetime of a defence. For example, in the face of sea level rise, the existence of coastal defences can contribute to 'coastal squeeze'. This is where inter-tidal habitats are unable to move landward due to the presence of sea defences and are gradually inundated by rising sea levels. Many of these intertidal habitats are designated for protection under the Habitats or Birds Directives.

Some impacts are associated with the maintenance phase of a defence, but these are generally considered to be less significant when compared to the original construction of a defence.

10.3 Moving from EIA to EMS

The regulatory context for EIA in flood risk management can be complex. Depending on the particular circumstances associated with a proposal, the EIA regulations under which the assessment is taken can vary because England and Wales have a number of sectoral EIA regulatory regimes. In some circumstances, it can appear that more than one EIA regime applies and agreements have to be reached with regulatory authorities as to which will lead on the consent for the project or on how their responsibilities will be coordinated. For example, coastal defences may have elements on the land that require planning consent from the local planning authority (local government) and elements below the mean high water spring tidal level that require a licence from the Marine Management Organisation. Both systems can have EIA requirements associated with them and a coordinated approach is required to ensure that the requirements of each consenting body are met. The regulatory context can be made more complex where sites protected under the European Habitats and Birds Directives are present. Demonstrating compliance with the European Water Framework

Directive is a requirement for all flood risk management projects. The objective of this Directive from the European Union is to achieve the sustainable use of water across Europe. Member states are required to protect and enhance the ecological and chemical status of all surface water bodies (rivers, lakes, coastal waters and estuaries) such that they meet good status by deadlines set in the Directive. Any new modifications to a water body are not permitted to cause a deterioration in the status or prevent the water body from reaching good status. Under exceptional circumstances causing a deterioration can be permitted, but only provided the strict conditions set out in the Directive can be demonstrated to have been met.[6] The assessment of compliance with the Water Framework Directive is incorporated into the EIA process. More positively, flood risk management can contribute measures that have been identified as being necessary to move a body of water toward good status. For example, the removal of weirs might have flood risk benefits and also reduce barriers to fish migration.

This regulatory complexity, coupled with the duties on the Environment Agency to protect and enhance the environment, requires the Agency to undertake active management of the environmental risks. The Environment Agency's National Environmental Assessment Service (NEAS) fulfils the role of environmental project managers on all projects where there are significant environmental concerns, including all of those where an EIA is required by regulation. Construction of a flood defence is undertaken by contractors working with the Environment Agency and other advisors. The relationship between EIA and EMS for flood risk management does not just concern the internal arrangements for the Environment Agency's own EMS, but also those of the contractors.

Contractors that work for the Environment Agency are included in a framework agreement. The agreement includes environmental requirements that define their role in the management of environmental effects and sets out broad requirements for their own EMSs. For example, their early involvement in the appraisal of a project is one of the obligations in the framework agreement. Other requirements include:

- Consulting with the Environment Agency's NEAS where methodologies

[6]Directive 2000/60/EC of the European Parliament and of the Council of 23 October 2000 establishing a framework for Community action in the field of water policy.

or on-site management deviate from that stated in the Environmental Statement or associated documents. This enables any legal implications to be checked and for the environmental implications of the changes to be assessed.

- Translation of mitigation and management measures set out in the Environmental Statement into contractual requirements for the project. These, in turn, are translated into environmental and site management systems and processes to ensure that they are delivered on site.
- Contractors having ISO 14001 certification in place during the life time of the framework agreement. The Environment Agency does not specify the scope of the EMS, but does set out key issues that are expected to be covered, these include:
  - Resource efficiency, in terms of the use of materials and minimising waste
  - The implementation of materials and site waste management plans
  - Avoidance of the use of virgin, finite resources where practicable and use of materials from recycled or renewable sources
  - Sourcing of materials locally where practicable
  - Identification of opportunities for delivering low carbon solutions (using the Environment Agency's carbon calculator to explore alternatives)
  - The purchase of timber from legal and sustainable sources and use of on-site borrow pits where practicable. Borrow pits are used as habitat creation opportunities.

Given that the key environmental aspects of a flood defence are primarily associated with construction, the management of the measures designed to minimise adverse effects is significantly reliant on the EMS of the contractor. The Environment Agency does not have direct control over this. Defining the relationship between EIA and EMS is not therefore about the seamless integration of the two instruments, but is focused on how issues are passed from the EIA into the on-site management requirements for the contractor. The detail of how we approach this is covered in section 10.3.1.

The Environment Agency does seek to provide assurance of the environmental performance of its contractors by monitoring on a site/project level and providing a more strategic overview. The Environment Agency maintains a sustainability scorecard. This monitors the performance across all projects against a range of sustainability factors. This can then be used to indicate the areas in which the contractors need to be influenced to improve overall performance. It also provides an indication of those aspects that are adequately managed by the contractor's own EMSs. The sustainability factors monitored are:

- Construction site safety
- Stakeholder relations (measured by the percentage of sites with a score of over 35 in the Considerate Constructors Scheme)[7]
- Reducing flood risk (percentage of houses protected relative to government targets)
- Using diverse teams (percentage of staff within organisations delivering schemes from black and minority ethnic groups)
- Including social enhancements in schemes
- Considering economic costs and benefits
- Linking with funding partners
- Creating new habitats
- Protecting endangered species (schemes benefiting Biodiversity Action Plan species)
- Minimising environmental incidents
- Reducing our carbon footprint
- Waste management (reducing waste going to landfill)
- Use of secondary or recycled materials
- Use of local contractors

The sustainability factors are designed to reflect wider measures of the performance of flood risk management at a programme level. These are adapted as the priorities set out by the Government change, whilst maintaining the spread across the factors that contribute to sustainability. For example, the current biodiversity priorities for the water environment now relate to the Water

[7]The Considerate Constructors Scheme is a scheme that commits contractors to be considerate and good neighbours, as well as respectful, environmentally conscious, responsible and accountable. In addition they must also consider their appearance and safety. Monitors visit a site to rate their performance against the criteria contained in a code of considerate practice. The maximum score that can be achieved is 40 points. Further information – `http://www.ccscheme.org.uk/`

Framework Directive and therefore it is possible that the sustainability scorecard will be adapted in the future to reflect this.

10.3.1 EIA: A Tool for On-Site Management

A product resulting from the EIA process is the Environmental Action Plan (EAP). This captures the mitigation and management measures identified during the EIA and set out in the environmental statement, covering the following project stages:

1. Pre-construction
2. During construction
3. Post-construction

The mitigation and management measures are set out in a table that also includes details of the responsible parties and any further actions may be required (Table 10.1 shows a typical format). The EAP is a 'fluid' document that is revised and updated as required during the detailed design and construction stages. Prior to the start of construction, the contractor is expected to translate the EAP actions and behaviours into their environmental and site management systems and processes that will ensure implementation of objectives on site and delivery of outputs.

The intent is that routine on-site environmental management (e.g. pollution prevention, waste management, etc.) is covered by the contractor's own EMS. The EAP focuses on those mitigation and management measures that are specific to the project and the sensitivities associated with the particular location. The actions in an EAP could cover one or all of the following:

- Actions to mitigate adverse effects identified in the Environmental Statement (or report) for the project.
- Actions to ensure that potential impacts are avoided.
- Actions to manage identified risks, such as those relating to potential damage as a result of increased vibration during scheme construction or the potential adverse effects on biodiversity or important features within the landscape.
- Actions relating to issues that were originally scoped out of the EIA but are considered to require some management during construction.

Table 10.1 *An example of the mitigation and management measures set out in the EAP; RSP = responsibility; FI = further information; M&O = monitoring and observation; FAR = further action required*

Ref.	Objective	Action	Target	RSP	FI	M&O	FAR
A. Prior to Construction							
Noise and Vibration							
A1.1	To minimise disturbance to residents of Road X as a result of increased noise levels during construction operations	Produce a noise monitoring and management plan and agree with the local authority and liaise with affected residents	No valid noise complaints	Project Manager			
Ground Conditions and Contamination							
A2.1	To minimise the mobilisation of contamination during piling works	Complete a Piling Risk Assessment Report once detailed design is completed and alter design and management of piling operations following recommendations	Understand any risks and identify recommendations and mitigation	Contractor			

Projects can take place in particularly sensitive locations and attract the attention of environmental groups (government and non-government), local government and/or the local community. In some instances the Environment Agency has established 'Environmental Task Groups' involving the various stakeholder groups, the contractor, other environmental advisors and, of course, expertise drawn from within the Environment Agency. The purpose is to communicate, discuss and adapt the steps to be taken to mitigate and manage the environmental issues, and provide confidence that the net effect of implementing a flood risk management project, if adverse, is not significantly so. The task group will often be established during the appraisal phase of a project and will continue to meet during implementation to verify that the environmental management measures have been implemented as agreed and to provide a means of validating any alterations to the environmental management approach where project changes or circumstances dictate.

The EAP sets out the actions to be taken to manage the environmental effects, but the Environment Agency takes steps to ensure that the selected contractor has the appropriate skills and experience to adequately implement the measures. Environmental requirements are included in the procurement process by:

- Setting environmental quality criteria and tender questions, preparing a performance specification and specifying other aspects of the Works Information. At this time, a determination will be made as to whether there is a requirement for an Environmental Clerk of Works (ECW) to supervise the construction works.
- Undertaking an evaluation of the tenders received in order to identify a preferred supplier – with particular emphasis on considering how the environmental requirements will be met by the contractor – especially where specialist environmental sub-contractors are required.

Once all necessary permissions have been obtained, a contractor has been appointed and the design has been sufficiently developed, work on the ground can begin. At this stage, the focus moves to implementing the EAP. The framework agreement sets out that the contractor should have full awareness of the contents and implications of the outputs of the EIA or SEA (strategic environmental assessment) process, the EAP and any detailed environmental drawings such as the landscape masterplan.

Contractors are involved in the early development of a project for the purpose of providing assurance that the proposed approach is feasible and providing an estimate of the likely costs. Nevertheless, when they take primary responsibility for the delivery of a project, design or methodology changes can result from 'ground truthing' the proposed approach. NEAS maintain oversight of any design, methodology or site management changes that significantly deviate from the design or approach stated in the Environmental Statement or Environmental Report (or on which planning consent has been granted). Changes are reviewed to determine whether they are significant. If they are likely to alter the conclusions of the Environmental Statement or affect planning conditions, it may be necessary to consult with consenting bodies to gain their agreement for the revised approach.

An ECW may monitor the construction process and will undertake periodic audits against the requirements of the EAP. Monitoring frequency will depend on the specific requirements of the project and the particular stage of construction. As actions are completed, these are signed off to validate that they have been appropriately implemented. However, it is important to note that some landscaping requirements may require ongoing monitoring over a five year period. In the event of an environmental incident occurring an investigation into the causes is undertaken and could result in the issue of a red or yellow card. This would trigger the requirement for the contractor to develop an action plan to address the management failure that lead to the incident. Ongoing performance will then be monitored for a period. Yellow cards act as warnings, whereas red cards will result in financial penalties for the contractor.

10.3.2 Sharing Lessons Learnt and Best Practice

For all projects undertaken by the Environment Agency, a project closure report is produced. This report documents how successful the project has been in delivering its intended outcomes. Within this report is a section on lessons learned, which is fed into other projects. One example of the success of this approach is on the Truro River, where works were undertaken on a floodwall along the edge of Lighterage Quay. Lessons learned from the Humber Estuary enabled the Environment Agency to deliver flood risk management obligations on the Truro River at a time when there were likely to be fewer conflicts with wintering birds' use of the area.

The works consisted of construction of a new 300m long quay wall 1m riverward of Lighterage Quay. This would result in encroachment into the area designated as a Special Area of Conservation (SAC) under the EU Habitats Directive and Site of Special Scientific Interest (SSSI) under national legislation, resulting in a loss of 300 m^2 of intertidal habitat. This impact was mitigated by the creation of intertidal habitat in advance of the works to the quay.

In total the works to Lighterage Quay took approximately 20 weeks to complete. Due to the condition of the existing flood walls the main works were planned to be undertaken in winter and there was a concern that noise could disturb any overwintering birds using the tidal mudflats in the Malpas Estuary SSSI. An assessment was made of the likely construction noise. This concluded that the total construction noise at a distance of 200 m from the source would be 68 dB(A). The range of steady noise levels that were observed in a Humber Institute of Estuarine and Coastal Studies Report (IECS)[8] to be tolerable (i.e. head turning but not taking flight) to birds located 200 m away from the source was 55 dB(A) to 85 dB(A).

Based on the findings of the IECS Report the level of noise expected from the proposed works could, therefore, result in head turning behaviour and reduced feeding movement to areas close to noisy activity, but this level of noise was not considered high enough to cause flight.

The following mitigation measures were then included to reduce the noise of the construction works and thus limit the impact on wintering birds:

- Movax vibration piling methods were used where possible to minimise noise levels generated. This was at the time the quietest known practicable piling method. Low noise generating plant was used.
- Noise monitoring was undertaken to ensure that noise levels did not exceed the levels referred to in the Humber IECS Report.

[8]Institute of Estuarine and Coastal Studies, University of Hull (2009), *Construction and Waterfowl: Defining Sensitivity, Response, Impacts and Guidance: Report to Humber Industry Nature Conservation Association.*

10.4 Achieving Continual Improvement

The Environment Agency approach to EIA and integrating the outcomes of this into the planning and implementation of a project are well understood by the framework contractors and consultants. There have been significant successes in reducing the environmental impact of our flood risk management activities. For example, the percentage of waste from our schemes going to landfill has been reduced from nearly 50 per cent in 2009–2010 to less than 10 per cent, just one year later. Nevertheless, there is always room for improvement both in the practical implementation of measures and the approach that sets the framework for improved environmental performance. The following identifies three challenges where improving our approach to EIA and EMS should add further value in terms of our environmental performance and the cost effectiveness of schemes.

10.4.1 Taking a 'What if' Approach to Risk Management

As previously stated risk management is a useful tool for ensuring successful delivery of environmental outcomes. For example, the risk of finding a protected species is often identified on the risk register at the planning stage of the project and that risk is followed through all stages from appraisal to construction. Generally the approach adopted is to undertake surveys to assess the presence of protected species. If present then this risk is managed through smart design or application of mitigation measures.

Although generally this approach is adequate it can lead to increased pressure on the construction programme and cause conflict with delivery. The challenge is therefore to determine whether taking a 'what if' approach early on in a project's life cycle will provide the same value in avoiding adverse environmental effects and simultaneously reduce the pressure on the project programme. For example, if we approached the early stages of planning a project by posing the question, 'what if protected species are found?' this would help us to start considering approaches to developing management measures at an earlier stage. It would inform the programme of works at an early stage and reduce the risk of programme delays and thus reduce costs. Other benefits would include early and productive dialogue with other interested conservation organisations including

Natural England and the Countryside Council for Wales.

10.4.2 Post Construction – Aftercare and Maintenance – Measuring Success

The Environment Agency can be confident in some of the environmental data generated as part of the construction programme. For example, the percentage of waste going to landfill and the carbon cost of a project are easily determined. There is less certainty over whether some of the longer term environmental and social measures have resulted in the benefits intended. Similar to other organisations, the monitoring of environmental outcomes following the delivery of a project tends to be under funded.

The lack of information on the success (or otherwise) of some measures can often prevent us from learning valuable lessons that would improve future schemes. There are examples where Environment Agency staff have undertaken informal site visits of their own volition. However this is ad hoc and a more systematic approach is required to realise the benefits of undertaking more routine post-construction monitoring. This would potentially lead to less corrective action on future schemes and result in a long term reduction in costs.

The requirements of the Water Framework Directive provide a particularly compelling reason to undertake post project monitoring. Undertaking flood risk management activities provides an opportunity to implement measures that will contribute to improving the status of a water body. However, without adequate monitoring it is difficult to be certain of the contribution that a single measure (e.g. habitat creation, improving river morphology, establishing reedbeds to improve water quality) will make to the improvement in status. As a result, it is more difficult to justify the additional expenditure this might require when the benefits are uncertain.

10.4.3 More Coherent Integrated Approach to the Application of EMS

Responsibility for environmental management for flood risk management is dispersed across various departments in the organisation. For example, those responsible for procurement assess the sustainability credentials of the suppliers and contractors. This can often involve simple consideration of accreditation to ISO 14001. Although this is undoubtedly important, it does not always allow consideration or assessment of particular skills or experience to address the environmental requirements associated with a particular scheme. NEAS manages the production of the EAP and contractor compliance with it using an ECW. Those responsible for delivering capital projects need to ensure that construction materials are re-used or recycled. All of these individual elements contribute to a coherent EMS. Nevertheless, these responsibilities can be addressed as separate entities rather than as a component that contributes to a coherent whole. The future challenge is to adequately integrate these components to provide an even more robust overall approach and improve standards further. At present it is difficult to assess quality, effectiveness and efficiency of effort. A centralised management approach would enable lessons to be learned more efficiently and improve the overall future application of an EMS.

11. Moving Towards Sustainability Management Systems

Jane Scanlon and Jenny Pope

11.1 Introduction

The relationships between environmental impact assessment (EIA) and environmental management systems (EMS) have been discussed extensively in the literature, often with a focus on ways to promote alignment between the two tools to ensure the effective management of environmental impacts of new developments from the approvals decision through to operation (Marshall, 2002; Sánchez and Hacking, 2002). Management systems, it has been argued, can play a key role by providing a framework to support the systematic management of project issues and risks (Ridgway, 2005).[1]

While we endorse the use of management systems to manage relevant issues throughout the life of a project, in this chapter we promote an approach based not upon traditional EMS but upon sustainability management systems (SMS),[2] an emerging concept reflecting a more holistic and societal-based consideration of issues, risks and opportunities inherent to project delivery. We ground our argument for SMS in our experiences as consultants and researchers on infrastructure projects, which include rail, road, water and power supplies (Gilpin, 2006), although we believe the principles we present may prove equally appli-

[1]From an impact assessment perspective, EMS is thus often conceptualised as a key element of 'EIA follow-up', the term given to a range of post (approval) decision activities (Arts et al., 2001).

[2]This acronym should not be confused with the acronym for safety management systems (also SMS) often used by occupational health and safety professionals.

cable to any project in the built environment.

Delivering sustainability outcomes in the built environment is a hot topic and increasingly on the agenda of project proponents,[3] their delivery partners, as well as the industry, government and the community at large. Sustainability assessment[4] techniques are being formulated to drive the selection and planning of projects that contribute to the sustainable development of society. Various organisations, including the Australian Green Infrastructure Council (AGIC),[5] and the Institute for Sustainable Infrastructure in the US[6] are undertaking activities that aim to advance project sustainability outcomes including through the development of sustainability-rating tools.[7] Such activity demonstrates a desire to re-think conventional project development practices so as to achieve sustainability outcomes across the life of a project.

However, guidance on managing sustainability in planning, design and construction activities has been lacking in the literature and within industry (Shelbourn et al., 2006). Academic research has typically focused on developing sets of indicators and frameworks for assessing or 'checking' sustainability (Shelbourn et al., 2006), while sustainability-rating tools, such as those outlined above, typically do not provide guidance on how to manage project development practices to achieve the performance outcomes that they assess.[8] In the last few

---

[3]Infrastructure proponents in various sectors, such as transport, include information on their organisational sustainability policy and agenda on their websites. See for example NSW Transport Construction Authority's Sustainability Targets available at www.tca.nsw.gov.au/Environment-and-Planning/Sustainability/default.aspx.

[4]Devuyst (1999, p. 461) defines sustainability assessment as 'a formal process of identifying, predicting and evaluating the potential impacts of an initiative (such as legislation, regulation, policy, plan, programme, or project) and its alternatives on the sustainable development of society'. However, the term 'sustainability assessment' is increasingly used more generally to mean a process 'designed to direct planning and decision making towards sustainable development' (Hacking and Guthrie, 2008, p. 73).

[5]See www.agic.net.au/.

[6]See www.sustainableinfrastructure.org/.

[7]Sustainability-rating tools use an assessment process to encourage and publicly recognise good practice across social and environmental (and sometimes economic) parameters in the development of projects in the built environment. Other examples include for buildings, BREEAM in the UK and LEED in the US; and for infrastructure, CEEQUAL from the UK Information on these tools can be found on the internet.

[8]In section 2.1.4 of the manual for the Institute for Sustainable Infrastructure draft sustainability-rating tool Envision, tool users are asked 'has a sustainability management system been established for the project commensurate with the level of complexity of the sustainability issues relevant

years some guidance on managing sustainability has started to emerge in other sectors and contexts. For example, British Standard 8901:2009 Sustainable Management System for Events (BS 8901) is a management system standard pertaining to events.[9]

Despite the lack of published guidance, various project proponents and their delivery partners are developing their own approaches to managing sustainability. In our experience, the transport sector is leading in this area, and two case studies in particular (the Sydney Metro Authority and the Olympic Delivery Authority Transport Division discussed in Chapter 12) have assisted in the development of our guidance below on SMS.

This chapter presents our proposal that an SMS should provide a framework to ensure that sustainability issues (including but not limited to those arising from formal impact assessment processes) are taken into consideration at the appropriate points in the life of the project by the appropriate people (decision makers). The sustainability implications of the different phases of a project are discussed in more detail in section 11.2.1. In section 11.2.2 we discuss what sustainability means for projects in the built environment, while section 11.2.3 considers some of the limitations of traditional EMS as a tool to support sustainability in project delivery.

In section 11.3, we offer our contribution to the emergence of SMS by providing some guidance on how SMS might be developed and implemented for projects in the built environment, for example in commercial buildings and civil infrastructure. While we believe it would be inappropriate to be prescriptive in how an organisation should establish an SMS (see also Boswell (2010)), because each project will vary in its regulatory, environmental and

to the project?' and 'have the appropriate mechanisms been put in place to manage the project's sustainability issues, impacts and opportunities?'. In the scoring module that provides further guidance on these questions, the 'Plan, Do, Check, Act' model is promoted, as we promote in this Chapter. However, currently the Institute does not provide significant guidance or case studies on establishing and implementing an SMS. The Institute for Sustainable Infrastructure's draft tool, comprising a number of documents including the manual and scoring module, is publicly available for comment and can be downloaded at www.sustainableinfrastructure.org/.

[9]BS 8901 is currently being developed into ISO 20121 Event Sustainability Management Systems. While these standards are targeted at event management, they are also relevant to developing an SMS for use in infrastructure delivery and operation activities. Such activities are usually required for major events such as the Olympic Games. As discussed in Chapter 12, the Olympic Delivery Authority Transport Division used BS 8901 in this manner.

social constraints and opportunities, there is still opportunity to provide high-level guidance on developing an SMS, in the manner of a standard such as ISO 14001 environmental management systems.

In summary then, this chapter provides the case for, and guidance on, developing SMS in the specific context of delivering and operating projects in the built environment based on the authors' experience, case studies of real-life projects, industry guidelines and standards, and the literature.

11.2 The Case for SMS

In this section we highlight the distinguishing features of an SMS to facilitate the delivery of project sustainability outcomes. We commence with a review of environmental and sustainability management throughout the life of a project, then consider the concept of project sustainability, before critically reviewing the potential of EMS and integrated management systems to deliver project sustainability.

11.2.1 Phases in the Life of a Project

The primary challenge considered by much of the literature on EIA and EMS is one of alignment, and how EIA, which is largely conducted for the purposes of appraisal and approvals (Sánchez and Hacking, 2002), can become an effective planning tool for the ongoing management of these potential environmental impacts throughout the life of a project. A potential disconnect arises because EMS is typically concerned with an organisation's ongoing activities, either at an operational or corporate strategic level, while EIA applies to proposed new activities or projects. The challenge, then, is how to ensure that environmental issues are managed throughout project delivery, before the project is operational (Marshall, 2002). This concern applies equally to the relationship between sustainability assessment and SMS.

Contributions to this literature generally either explicitly or implicitly recognise that project delivery is a complex process comprised of many inter-related and iterative activities, most if not all of which have the potential to cause impacts on the environment. Due to this complexity and the reality that different activities at different stages of the project life are often managed by different groups of people, there is considerable risk of issues 'falling through the gaps', to not being appropriately addressed at the appropriate point by the appropriate decision makers. In considering the life of a project, we will refer to the following categorisation, derived from Ridgway (2005), throughout this chapter:

- Planning (feasibility, project definition, concept design, approvals);
- Design (detailed design and documentation);
- Construction (procurement, construction and commissioning);
- Operations (operation, maintenance and decommissioning).

Our proposal for SMS, outlined in section 11.3, encompasses activities within the project delivery (design and construction) phases, between approvals and operation.

In the literature, there is some discussion of the links between EIA and EMS throughout project delivery. For example, Marshall (2002) suggests that project environmental management plans (project EMPs) should provide the crucial link between EIA and EMS through the design and construction phases of a project. Since it is often the case that only very preliminary consideration is given to design options and construction methods for the purposes of EIA (Ridgway, 2005) and therefore commitments made in this regard are often vague (Sánchez and Hacking, 2002), it is unlikely that an EIA will identify all potential impacts relevant to design and construction. Furthermore, designs and construction methods are likely to evolve significantly throughout the project and as the project detail develops. The project EMP must therefore be flexible enough to respond to these changes. It should also be developed in collaboration with delivery partners responsible for these project phases, particularly as they may not have been involved at the planning phase at which the EIA was undertaken. These principles of flexibility and engagement are also essential components of our proposal for an SMS (see section 11.3).

Procurement should also be explicitly included in project EMP, since an organisation's corporate EMS may not be focused on driving sustainable project procurement. Lam et al. (2011, p. 790) highlight this, stating 'EMS has little influence on the contractual issues which play important roles on the green performance of projects'. They add that even if the contractor is complying with its EMS, the project 'may still have negative environmental impacts because specified materials and practices in the contracts are not conducive to sustainable construction'.[10]

The importance of integrating environmental or sustainability considerations into design is discussed by various commentators including Ugwu et al. (2006), Vanegas (2003) and Spangenberg et al. (2010). For example, Vanegas (2003, p. 5368) states that early delivery phases have the 'greatest potential to influence overall project sustainability at the lowest cost' because it is during these that the shape and form of a project is being determined. It is at the design phase that significant decisions are made that affect sustainability. This includes in the specification of construction materials and their environmental, social and economic performance from a whole-of-life perspective.

Despite the recognition of the importance of integrating environmental or sustainability considerations into different activities within the various project phases, what appears to be lacking is one uniting framework that ensures that all relevant issues are consistently managed, at the appropriate time and by the appropriate decision makers. The lack of guidance is problematic because integrating sustainability into the day-to-day activities and decision making of an individual or discipline must be actively managed. If it is not managed, those individuals will likely resort to standard practice, which is not enough to achieve sustainability outcomes.

While much of the previous discussion has been concerned with environmental management through the life of a project, the same issues are relevant when considering project sustainability. Furthermore, sustainability brings with it additional challenges, some of which are discussed in the following section.

[10] While this chapter does not go into detail about green/sustainable procurement, Lam et al. (2011) discuss the need for 'green specifications', which are a set of contractual requirements addressing sustainability, to both translate sustainability theory into construction practice, and to improve on the capability of EMSs. Procurement in this context is discussed further in Chapter 12. Also, guidance on sustainable procurement is increasing and is available from Australian Green Infrastructure Council (2011) and British Standards Institute (2010) Principles and Framework for Procuring Sustainably.

11.2.2 Project Sustainability

While it is clear that some projects may be more 'sustainable' than others – for example a renewable energy compared to coal-fired power station project – the delivery and subsequent operation of a project requires explicit consideration of sustainability. Sustainability in the context of delivering a project is about risks and opportunities identification and management (Scanlon and Davis, 2011; Spangenberg et al., 2010). Sustainability risks and opportunities will span across environmental and social (including socio-economic) dimensions. The Australian Green Infrastructure Council (AGIC), for example, in its pilot sustainability-rating tool, encourages and recognises sustainability performance under themes or categories such as: project management and governance; economic performance; using resources; emissions, pollution and waste; biodiversity; people and place; and workforce (Australian Green Infrastructure Council, 2009).

Sustainability opportunities might include creating ongoing local employment, improving housing affordability and enhancing community relations through creating green space. Risks might include future liabilities in an increasingly carbon constrained economy, and structural design risks associated with climate change. Risk and opportunities identification and management are good but by no means standard practice. However, project sustainability will also require effort to go beyond legislative compliance, and adopt innovative, creative thinking and continual improvement in response to ever-changing political, technological, cultural, economic and environmental contexts; and doing so in a technically viable and cost effective way.

Trends in impact assessment practices to incorporate sustainability considerations both in project planning and in the regulatory approvals process – see for example Pope (2006); Pope and Grace (2006); Gibson et al. (2005); Hacking and Guthrie (2008) – mean that there is often considerable information about the social and sustainability impacts of a proposed project at the conclusion of the planning phase. Some of this information might be incorporated into approvals documentation as the basis of commitments that subsequently become legally binding conditions. Other relevant sustainability information that may be outside the regulatory approvals process, including the views of community and stakeholders obtained through engagement processes during project planning, also needs to be factored into subsequent decision making processes. The challenge is then to ensure that all these issues are adequately addressed in subsequent project phases. In the context of this chapter, the question then becomes: 'to what extent can EMS (and other management systems) provide the framework for sustainability management within projects in the built environment?'

11.2.3 EMS, IMS and Sustainability

A typical EMS is generally (but not necessarily) based on ISO 14001, which is part of the ISO 14000 environmental quality standards. In theory, EMS is a tool for managing the environmental dimension of sustainability. Moving beyond a purely environmental focus, integrated management systems are a promising approach to project sustainability. An integrated management system is a management system that holds under one 'banner' or framework various other systems, plans and procedures – for example occupational health and safety (OH&S), human resources, financial control and environmental management. Various standards and guidelines exist for integrated management systems.[11]

Benefits associated with integrated management systems have been discussed by commentators including Johnson and Walck (2004) and Oskarsson and von Malmborg (2005). They include potential for improved internal collaboration, engagement and participation; and streamlined auditing and management review processes (Johnson and Walck, 2004). Integrated management systems can also help reduce system duplication and complexity (AS/NZS 45801:1999).

Integrated management systems can contribute to sustainability thinking. This is because they can incorporate information about a broad spectrum of issues, all of which need to be viewed concurrently, not in isolation. Therefore, they can encourage more integrated and less 'siloed' thinking and decision making, as well as highlighting trade-offs between and between the different dimensions.[12] However, in practice, most integrated management systems tend to be internally focused and to reflect more traditional topics such as quality,

[11] For example the Australian/New Zealand Standard AS/NZS 4581:1999 Management System Integration – Guidance to Business, Government and Community Organisations.

[12] It can also be argued that adapting EMS to incorporate other sustainability dimensions may be inappropriate because the focus on managing environmental impacts may be lost to social – but more likely economic – imperatives.

OH&S and environment – see for example (Jørgensen et al., 2006) – rather than address broader sustainability risks and opportunities that reflect the impact of the project on society, thus limiting their potential to facilitate sustainable outcomes.

There is some literature on the potential to expand EMS to incorporate social dimensions in the form of a sustainability management approach, although not in the specific context of projects (Emilsson and Hjelm, 2009; Palframan et al., 2010; Sarkis and Stroufe, 2007). In addition, some companies are specifically incorporating corporate social responsibility commitments into their integrated management systems, moving them closer towards addressing sustainability (Salomone, 2008). Despite the potential of integrated management systems, particularly those explicitly incorporating the management of social issues, the underlying philosophy of management systems as promoted by ISO is somewhat inconsistent with some aspects of sustainability as we perceive it. Essentially, traditional management systems are about minimising risks, with each organisation establishing its own level of performance beyond minimal compliance with legal and other requirements (Johnson and Walck, 2004). There is no requirement to develop objectives that reflect opportunities or positive sustainability outcomes, only ones that demonstrate continual improvement in minimising negative impacts (MacDonald, 2005).

It can also be argued that ISO-compliant management systems provide little incentive for an organisation to move significantly beyond legal compliance. A focus on legislated targets tends to restrict or discourage innovation (Carruthers and Tinning, 2001), which we believe are essential to sustainability leadership in project delivery. We suggest that this innovation and creativity is promoted by a multi-disciplinary work environment, in which the sustainability (and environmental, OH&S, and community professionals) interact with those within the core project delivery function, including project managers, designers, procurement professionals and others. This interaction is essential for ensuring the sustainability opportunities are realised in a cost effective and technically viable way. A traditional management approach typically does not encourage or have provision for this necessary level of multi-disciplinary interaction.

Furthermore, community and stakeholder considerations do not feature in a typical EMS at all – there is no requirement for public participation beyond the requirement that the Environmental Policy should be made public (Eccleston and Smythe, 2002). We believe that engagement processes are similarly essential to sustainability, to ensure that broader societal values, views and knowledge are incorporated into decision making. Such engagement should commence in project planning, as part of the EIA or sustainability assessment process, and continue throughout the life of the project.

11.3 Developing and Implementing an SMS

From the discussion in section 11.2, we can distill the characteristics of an effective SMS, which differentiate it from EMS or ISO-compliant management systems generally:

- The SMS should provide a framework for purposeful and planned integration of sustainability considerations through all phases of the project life, particularly during design and construction phases;
- The SMS should address sustainability (not just environmental) issues identified during project planning and impact assessment processes, including but not limited to issues that are explicitly included in project approvals;
- As such, the SMS should be sufficiently flexible to incorporate sustainability issues, including environmental and social risks and opportunities, that emerge during project delivery, and to facilitate innovation and creativity;
- The SMS should move beyond mitigating negative impacts and risks to achieving positive outcomes and realising social and environmental opportunities;
- The SMS should incorporate key sustainability characteristics including long-term thinking, multi-disciplinary interaction and decision making, and community and stakeholder engagement.

We do not propose that an SMS should replace traditional management systems such as for quality, OH&S, human resources or even for environment. Other management systems are well established and have an important function within their respective spheres of influence. Rather, an SMS will allow for the management of project sustainability, as described above, in a manner that other management system approaches currently do not have the capacity to facilitate. However, a critical feature of an SMS is that it is integrated with these other management systems, plans and procedures given the inter-disciplinary nature of the sustainability concept (Figure 11.1). How this integration can be achieved is discussed further below. We propose that an SMS be distinct from other management systems so as to ensure that sustainability activities are not 'absorbed' into other systems, plans and procedures where they might become 'lost', for example 'designed out' in the name of simplifying the construction process (Pulaski and Horman, 2005) – see also Chapter 12.

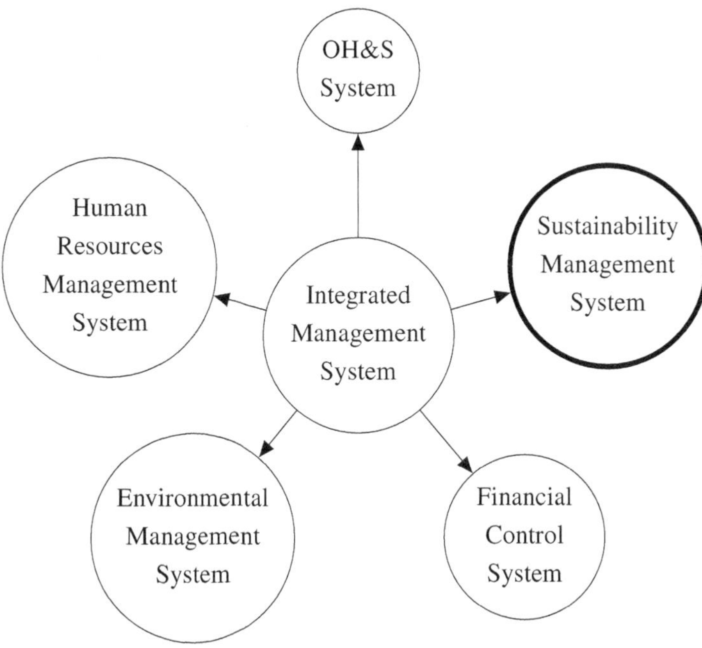

Figure 11.1 SMS within an integrated management system structure

Other benefits of having an explicit SMS include being able to document and therefore demonstrate sustainability commitment leading to improved reputation, and even (particularly for contractors) enhanced commercial opportunities and competitive advantage. An SMS can also assist in overcoming a number of inhibitors to project sustainability, as discussed in Chapter 12.

In this section, we offer some practical guidance on developing and implementing such an SMS. The SMS model we propose is based on the 'Plan, Do,

Check, Act' (PDCA) model that forms the basis of other management systems, including ISO 14001:2004. According to the American Society for Quality (2011), 'PDCA is a four-step model for carrying out change. Just as a circle has no end, the PDCA cycle should be repeated to assure continuous improvement' by feeding back into the SMS.

11.3.1 Plan

The 'Plan' element of the PDCA model sets the scene for sustainability for the rest of the management system. Planning activities include establishing a policy commitment, and developing and prioritising objectives, targets and initiatives. These steps should be informed by the sustainability issues identified during the project planning phase, including the EIA and/or sustainability assessment and community and stakeholder engagement processes, particularly those for which specific commitments have been made or which approval conditions apply. It is also important to recognise, however, that the SMS should be sufficiently flexible to accommodate further sustainability issues that may emerge as the project progresses and evolves.

Establishing policy commitment; and objective, targets and initiatives
At this stage, policy commitments are developed to meet the vision and values for the project, demonstrating the social, environment and economic ideals for a particular project. An actual policy document (which can be called a 'sustainability policy') should be produced that is made known to every project-related employee and endorsed by the leadership team.

Specific sustainability objectives and targets[13] should be established in order to meet the policy commitments, and to realise the vision and values for the project. Objectives can be determined through risks and opportunities analysis, including via the use of standards such as AS/NSZ ISO 31000:2009 Risk Management – Principles and Guidelines – see also Fernández-Sánchez and Rodríguez-López (2010). Example sustainability objectives include: building the skills and capacity of local business; striving for a neutral or negligible

[13]In contrast, with traditional EMS, objectives and targets are focused on continual improvement in minimising negative impacts and risks.

construction carbon footprint; and achieving a net gain in biodiversity outcomes. Sustainability objectives will usually be reflected in the policy document.

Sustainability targets can be established through benchmarking of similar projects internationally and nationally that developed specific targets for themes such as using resources and biodiversity. Through benchmarking, it is possible to ascertain what targets were met on these other projects, and hence what might be achievable for the project in question. Example sustainability targets include using 30 per cent renewable energy during construction, and reducing potable water use during operations by 80 per cent against a 'base case' or standard project. Other examples of both sustainability objectives and targets can be found in Chapter 12.

In meeting sustainability objectives and targets, specific initiatives need to be devised, and indicators established or selected to measure progress toward those objectives and targets (although the actual measurement happens at 'Check' below). The practicality of the targets and initiatives should be demonstrated through feasibility and other studies (including benchmarking), as well as through engagement with stakeholders, including across the supply chain.

Prioritisation and materiality A project proponent may choose to prioritise sustainability objectives and targets, and set high targets for where there is greatest capacity to achieve them (British Standards Institute, 2010; Fernández-Sánchez and Rodríguez-López, 2010). The Olympic Delivery Authority Transport Division case study in Chapter 12 gives an example of this prioritisation. The concept of prioritisation is similar to that of materiality, the approach now being championed by the Global Reporting Initiative (2011) to ensure that an organisation is concentrating on sustainability issues most relevant to its operations. GRI promotes a qualitative assessment based upon a materiality quadrant which has on one axis 'influence' and on the other axis 'impact' as shown in Figure 11.2. This assessment must involve active participation and input from relevant internal and external stakeholders, and this involvement of stakeholders is a key point of differentiation between our proposed SMS model and a standard EMS. The materiality assessment should also reflect the findings of the EIA/sustainability assessment where available.

Prioritisation is important in preventing a 'shopping list' approach whereby every possible sustainability risk and opportunity is identified for management,

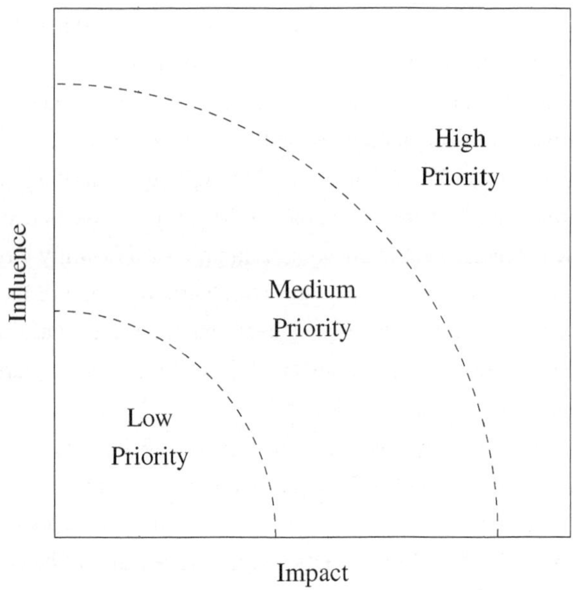

Figure 11.2 The materiality quadrant – adapted from the Global Reporting Initiative (2011).

each with specific objectives and targets. A shopping list approach can often overwhelm members of the project team, including the leadership team, and can result in disinterest in sustainability. The case for addressing specific sustainability issues therefore needs to be sound, and this case can be built on a materiality exercise, as well as by demonstrating the feasibility and benefits of meeting certain targets and adopting certain initiatives, including improved operational efficiency.

Integration As highlighted above, a key success factor for an SMS is the extent to which it is integrated into other project and organisational processes, rather than simply being an add-on to the organisation's way of doing business. For an SMS to be truly integrated there must be clear links with and effective take up in other systems, plans and procedures.

In the interests of integration, each objective, target and initiative needs to have a 'home' within other systems, plans and procedures, and these should be identified during the Plan stage. For example, an objective relating to reducing greenhouse gas emissions may have a target attached to it of reducing 25 per cent of energy use during construction against a base or business as usual case. Specific initiatives might include servicing construction plant and equipment to enhance its fuel efficiency, and investing in on-site renewable energy generation. A further initiative could be the monitoring of energy use on-site to ensure the target is being achieved, accompanied by the development of an energy monitoring pro-forma or template. This objective, target and associated initiatives should be reflected in the construction EMP and procedures. It should also be reflected in the procurement documentation – namely in ensuring it is written into request for proposal documentation so that construction contractors can price for it accordingly, and so that it can be made a contractual requirement.

11.3.2 Do

'Do' is about implementing the initiatives identified in the 'Plan' element to achieve the objectives and targets. This includes defining organisational structures and related roles and responsibilities; providing training, development and awareness; engagement, collaboration and participation by internal stakeholders; and engaging external stakeholders.

Organisational structure, and roles and responsibilities Broadly speaking, the term organisational structure refers to the framework that governs, coordinates and generates the activities of an organisation. The Office of the Commissioner for Environmental Sustainability – Organisational Structures and Sustainability (2011, p. 1) considers that organisational structures 'play an important role in embedding sustainability within an organisation by supporting work to sustain organisational change in the long term, and in facilitating the organisational cultural change that is necessary for sustainability'.[14] Best or leading practice is where every employee, including the leadership team, has clearly defined roles and responsibilities for sustainability, coupled with adequate authority and resources. Leadership team commitment and involvement is particularly important in achieving sustainability, as discussed by Johnson

[14] The importance of clearly defined roles and responsibilities is equally emphasised in ISO 14001:2004.

and Walck (2004), and further in Chapter 12.

Project proponents and their delivery partners are increasingly engaging sustainability advisers or consultants to address project sustainability (Scanlon and Davis, 2011) – see also Chapter 12. Sustainability advisers will be instrumental in implementing an SMS, and must have clearly defined roles and responsibilities in this regard. Sustainability advisers should also be capable of listening to, educating and mentoring project members to build their capacity to understand sustainability (Scanlon and Davis, 2011). Those individuals should also have access to the right skills and expertise to achieve the various sustainability objectives and targets. This will require a multi-disciplinary team working together to achieve sustainability outcomes.

Training, development and awareness As well as defining the roles and responsibilities of individuals, it is equally as important to train individuals in their particular role and to make them aware of what their responsibilities entail. Therefore training and development on sustainability should be provided, including 'refresher' training offered on a regular basis. Providing training, development and awareness is also about creating a 'culture' for sustainability. The notion of 'culture' in the context of sustainability has been reported on extensively in the literature, mainly in the area of corporate or organisational responsibility and sustainability – see for example: Harris and Khare (2002); Dunphy et al. (2003); Benn et al. (2006); Johnson and Walck (2004); Doppelt (2003).

Multi-disciplinary interaction The delivery and operation of projects involves multiple disciplines and work streams that have traditionally operated in isolation or in 'silos'. This compartmentalisation can be amplified by the type of contractual arrangement – for example in a design and construct contract compared to an alliance or other relationship contract (Scanlon and Davis, 2011). But because of the integrated nature and long-term implications of the sustainability concept, high levels of engagement, collaboration and participation amongst these different disciplines are required, and should be encouraged by the SMS. One way this can be achieved is by allocating certain sustainability targets to multiple individuals and disciplines, encouraging those individuals to work together to achieve those targets, which can have other benefits including team and relationship building. This focus on multi-disciplinary interaction is a distinctive difference between traditional EMS and an SMS as described here.

External stakeholder engagement Constant and appropriate stakeholder engagement is important throughout the entire life of a project. Such stakeholders may include regulatory agencies and community groups. Community members and other stakeholders can potentially be actively involved in the SMS through contributing to the prioritisation process (as discussed in section 11.3.1); helping to establish objectives, targets and initiatives; and participating in decision making, for example by participating in a multi-criteria assessment[15] process to determine the most sustainable location or design option (e.g. by providing inputting into the weighting process through which the relative importance of a range of factors influencing the decision are determined). Meaningful community and stakeholder engagement, through which external parties have an opportunity to influence decision making, can not only build good relationships but can also contribute to concrete sustainability outcomes through the infusion of fresh ideas.

11.3.3 Check

'Check' is about measuring performance in relation to the sustainability objectives and targets that were identified in 'Plan' and reporting these results so that data is made available to inform decision making and changes and/or improvements can be made.

Measuring and monitoring As discussed in section 11.3.1, our SMS model promotes prioritisation of sustainability objectives. Hence, sustainability indicators should be selected that allow for progress against prioritised objectives to be monitored and considered (and hence information on those issues can be fed back into the management system). Carruthers and Tinning (2001) also advocate the selection and use of indicators that are most relevant to the undertaking in

[15]Guidance on multi-criteria assessment can be found in Communities and Local Government (2011) *MultiCriteria Analysis: A Manual*, available at `www.communities.gov.uk`, accessed 5 February 2011.

question, rather than the use of a prescribed set of indicators. Further, because the most prevalent or material sustainability issues for a project will change over time due to a range of factors such as changes in community expectations, indicators should be altered to reflect this (Shen et al., 2007).[16]

Verification and assurance Verification and providing assurance are critical to the 'Check' element. Verification relates to checking whether the measured or claimed performance is accurate, and assurance pertains to demonstrating to internal and external stakeholders the outputs of – and appropriate action undertaken as a result of – that verification process. A strong verification and assurance process is crucial to check that organisations deliver on their sustainability commitments and requirements. Internal and external auditing processes also play a key role in providing assurance, as they do in traditional ISO management systems.

Reporting Once indicators have been measured and results have been collated, these need to be arranged into reports. These reports should draw conclusions and make recommendations for changes and/or improvements. The reports should be easy to understand and act upon. It is also important that results are collated and reported in a manner that allows them to be compared over time, and that reporting is undertaken on a regular basis.[17]

11.3.4 Act

'Act' is about intervening where necessary to ensure that the SMS is efficient and effective in producing the desired outcomes. During the 'Act' element, results that are produced in the 'Check' element are considered. These results are then linked back to the sustainability objectives and targets that were identified in the 'Plan' element to facilitate necessary change and/or improvement. This might include revising a particular target, or adopting a new initiative that has become both viable and desirable due to, for example, changes in technology or product availability. In an ideal situation, new objectives and targets should be

[16] The Global Reporting Initative (GRI) provides sets of indicators for measuring and reporting on sustainability performance (Global Reporting Initiative, 2011). GRI is being made more relevant to the built environment through the Real Estate and Construction Sector supplement.

[17] As mentioned above, GRI provides a framework for producing sustainability reports.

set higher as existing objectives and targets are met. These activities make 'Act' the cornerstone for continuous improvement – a management system principle that is consistent with the notion that sustainability is not a destination, but rather is a journey.

A critical component of 'Act' is leadership-level or management review, which should be undertaken at regular intervals (Tan et al., 2011; British Standards Institute, 2010), as is the case for traditional ISO management systems. Management reviews assess the suitability, adequacy and effectiveness of the system. A 'lessons learnt' programme can also be useful in feeding information back into the system. The importance of capturing lessons learnt in general, that is not specific to sustainability, is discussed by Kaioglou et al. (2000).

11.4 Conclusion

While both EIA and EMS have contributed enormously to good environmental outcomes in project delivery and organisational practices respectively, we have argued that standard practice as we see it today falls short of ensuring that sustainability considerations are fully integrated throughout the life of a project to achieve positive sustainability outcomes in the built environment.

Sustainability requires not just the systematic approach to ensuring legal compliance and incremental improvement offered by traditional management systems, but innovation, creativity, and multi-disciplinarity supported by broad community and stakeholder engagement. We have offered a model for an SMS that incorporates these characteristics, based upon the 'Plan, Do, Check, Act' model. Our concept of an SMS is illustrated in Chapter 12 through two case study SMSs from the transport sector – the Sydney Metro Authority in Sydney and the UK Olympic Delivery Authority Transport Division.

We suggest that such an SMS should not necessarily replace other frameworks and systems such as EMS but rather supplement them. While it could be argued that sustainability management does not require a separate (albeit integrated) approach or management system – that it should be inherent in all systems and processes thus avoiding the creation of yet another plan or system, we believe that a dedicated SMS, led by sustainability advisers, has the greatest potential to drive change beyond business as usual towards sustainability.

12. Sustainability Management Systems – Two Case Studies

Jane Scanlon, Hudson Worsley and Neil Earnshaw

12.1 Aim and Background

12.1.1 Contribution to this Book

This chapter links to Chapter 11, which defines sustainability in the context of delivering projects in the built environment. It also promotes the use of sustainability management systems (SMSs) across the life of a project and provides a high-level model for the development of an SMS. SMSs can ensure sustainability issues, and risks and opportunities, are meaningfully addressed during various project delivery activities. These activities include those pertaining to the design and construction phases, such as detailed design and procurement (see Chapter 11). Such activities, although critical for achieving sustainability outcomes, are often not covered by a traditional environmental management system (EMS).

This chapter outlines two real-life SMSs being used in the transport sector – one by the Sydney Metro Authority (SMA) in Australia and the other by the UK Olympic Delivery Authority Transport Division (ODAT).[1] It describes the key components of each SMS and how they have served to embed sustainability in various project delivery activities. This chapter also highlights how an SMS

[1]The case studies formed part of PhD research conducted by the primary author between 2008–2011 at the University of Western Sydney, Australia.

can overcome some of the inhibitors to achieving project sustainability. This chapter will be of particular interest to industry professionals who wish to drive sustainability through project delivery. Those professionals will include sustainability advisers or consultants[2] and impact assessment practitioners.

12.2 Case Study Backgrounds

12.2.1 The Sydney Metro Authority

The Sydney Metro project, that was being delivered by SMA, was a new underground metro system for Sydney, New South Wales. The project was deferred by the State Government in February 2010. At the time of its deferral, stage 1 had planning approval, and stage 2 planning and design was underway. Stage 1 comprised a seven kilometre long metro railway through Sydney's Central Business District (CBD) to the suburb of Rozelle.[3] Stage 2 was the first extension to stage 1, comprising a 24.1 kilometre long metro railway reaching to Sydney's western suburbs including Parramatta[4] and the inner west (Transport NSW, 2010a).

SMA had received proposals back from the private sector for delivery and operation of stage 1 at the time of its deferment. Despite its deferral and contest around the ability of the metro system to solve Sydney's public transport issues (Independent Public Inquiry, 2010), there is general consensus amongst industry professionals in New South Wales that the project was leading in actual and planned sustainability activities (Scanlon and Hodgson, 2010). It is therefore a good case study for examining the role of an SMS, and for sharing lessons learnt to the broader industry in legacy.

---

[2]The sustainability advisory profession is a professional service that offers advice on a variety of sustainability issues (such as life-cycle assessment, greenhouse gas assessments, and sustainable procurement) to project proponents, their delivery partners (such as construction contractors) and operators. In Australia, the number of individuals and companies claiming to offer such services are increasing. A report based on a survey, titled The 2011 Sustainability Roles and Salary Survey (`www.turninggreen.com.au/`), provides more detail on the sustainability advisory profession in Australia.

[3]Rozelle is located approximately 4 km west of the Sydney CBD.

[4]Parramatta is located approximately 24 km west of the Sydney CBD.

12.2.2 The Olympic Delivery Authority Transport Division

In July 2005, the International Olympic Committee chose London to host the 2012 Summer Olympic Games (the Games). The London Organising Committee of the Olympic Games and Paralympic Games (LOCOG) was formed soon after as the organisation responsible for the planning and staging of the Games. LOCOG works closely with the Olympic Delivery Authority, which has the primary role of developing and constructing the required infrastructure and venues (LOCOG, 2011a). The Olympic Delivery Authority's mission is to 'deliver venues, facilities and infrastructure, and transport on time and in a way that maximises the delivery of a sustainable legacy within the available budget' (Olympic Delivery Authority, 2007, p. 5).

ODAT[5] has statutory obligations under the Olympic and Paralympic Games Act 2006 for the movement of people during the Games. This has required delivering new and upgrading existing transport infrastructure. Projects have included trebling the capacity of Stratford Regional Station through activities such as widening station platforms and adding nine new lifts, and delivering the Orient Way project, a new 12 road carriage berthing siding (Olympic Delivery Authority, 2010). Other projects include the West Ham station upgrade, which comprises temporary and permanent legacy works including a new footbridge, sets of stairs, re-graded accessible ramp and a walkway to provide a link for passengers from the platform to the Greenway walking and cycling route in the Olympic Park (Olympic Delivery Authority, 2010). All of these projects have now been delivered.

12.2.3 Case Study Data

Semi-structured interviews[6] and a review of organisational and project-related (such as procurement) documentation were the main sources of case study data. Quotes from interviews with individuals from the respective case study organisations are used to highlight the discussion. Further, one of the author's

[5]ODAT operates relatively autonomously from the ODA, but the Transport Director is one of eight Olympic Delivery Authority directors. ODAT works with delivery partners including Transport for London, the government body responsible for most aspects of the transport system in Greater London, and the Highways Agency.

[6]Interviews were conducted by the primary author under PhD Ethics approval number H7609 from the Human Ethics Committee at the University of Western Sydney.

practical experience working as Sustainability Manager for ODAT and LOCOG is drawn upon. Another author's practical experience working as Environmental Sustainability Manager at SMA is also used.

For SMA a total of 12 and for ODAT a total of nine interviewees participated in the research. These interviewees were separated into four disciplines or work streams: environment and planning; sustainability; design; and leadership-level. SMA interviewees comprised three individuals in environment and planning (SE1–3); four in sustainability (SS1–4); two in design (SD1, SD2); and three from the leadership team (SL1–SL3). ODAT interviewees comprised one individual in environment and planning (OE1); three in sustainability (OS1–OS3); two in design (OD1, OD2) and three from the leadership team (OL1–OL3). The overall selection protocol was that potential interview participants had to either have been an employee of SMA or ODAT, or a delivery partner such as a construction contractor, and in some way have been involved in sustainability.

During the interviews, the journey in developing and implementing the respective SMSs was discussed, including how those SMSs served to overcome inhibitors to project sustainability. The interviewee responses were supported by various related documents detailing the SMSs, such as management system procedures.

12.3 Sustainability Management Systems

This section outlines SMA and ODAT's SMS. It separates the key components of each SMS into the 'Plan, Do, Check, Act' management system approach championed by the International Organisation for Standardisation, and in Chapter 11.

12.3.1 Sydney Metro Authority's SMS

Plan – Sustainability Policy, Themes, Objectives and Targets SMA's overall sustainability management approach was encompassed by its SMS. A Sustainability Policy formed the central sustainability governance document for the organisation and for all of its projects. The Sustainability Policy contained ten sustainability 'themes' of governance; energy; greenhouse gas emissions; climate change adaptation; resource efficiency; water; biodiversity; community and stakeholder involvement; community benefit; and economic vitality and viability.

A Sustainability Position Paper nominated policy positions on the different sustainability themes across a range from leadership to beyond compliance, and lastly to compliance-based. A policy position on each of the ten themes was determined based on an analysis of applicable government policies and legislation, and impacts and opportunities across environmental, social and economic dimensions. Numerous sustainability objectives were devised for each theme (Transport NSW, 2009a, 2010a) – for example, for:

- Governance – adhere to the SMS
- Energy – minimise energy demand, and use renewable energy
- Greenhouse gas emissions – achieve low emissions during construction
- Climate change adaptation – design to withstand the effects of climate change
- Resource efficiency – reduce waste generation
- Water – minimise use of potable water
- Biodiversity – achieve a net increase in biodiversity
- Community and stakeholder involvement – maximise access and connectivity
- Community benefit – facilitate healthy living in live-able communities
- Economic vitality and viability – create development opportunities around the metro corridor

Specific targets and initiatives for each objective were negotiated during concept design and procurement activities. A Sustainability Initiatives Register was developed by the design team in consultation with sustainability advisers. The Sustainability Initiatives Register included specific initiatives, for example for greenhouse gas emissions sourcing 20 per cent of electrical energy from renewable sources during construction, the use of energy-efficient site lighting, and the consideration of the embodied energy of materials such as cement and steel (Transport NSW, 2010a).

The Sustainability Initiatives Register was supported by various sustainability working papers that included technical, cost benefit and other information to inform decision making and implementation. This included benchmarking internationally to see what similar projects were able to achieve. For example, a working paper was produced on low carbon and renewable energy for operations. The sustainability advisers produced the paper which included an analysis of energy options, such as building a wind farm, for operation of stage 1 (CBD Metro project).

The development of the themes, objectives and targets took place at an early design stage when the reference design – the technical blue-print of the project – was being produced. Environmental impact assessments were also being undertaken around this time – namely for stage 1 and stage 2 as discussed above. The sustainability objectives were not explicitly part of the early business case or pre-feasibility stage of the project where funding was being sought. However, it was inherent to the vision and value of the project, because it was a metro system aiming to provide a public transport solution for Sydney – a city with significant transport issues (Independent Public Inquiry, 2010). The SMS was intended to maximise the inherent sustainability value of the Sydney Metro project, or as interviewee SS1 pointed out, 'the executive ... understood that one of the reasons you put in a metro is for sustainability. They also recognised that that doesn't happen unless you drive it in project delivery'.

Plan and Do – Integration across Delivery Activities There were three key delivery activities requiring integration of the sustainability objectives, targets and initiatives – namely design, planning approvals and procurement. Interviewee SL1 stated that '[SMA] pushed sustainability as an objective all the way through [its] documents'.

Integration into design Interviewee SL1 expressed the importance of sustainability in design, stating that 'you have to get [sustainability] in design right from the start. You have got to capture it upfront'. Vanegas (2003), Ugwu et al. (2006) and Spangenberg et al. (2010) also discuss the importance of embedding sustainability in design. If sustainability initiatives are locked into design drawings and specifications, and hence into contractual arrangements, the construction contractors will have to deliver them. It also means that tendering contractors can build initiatives in as part of their proposed fees, and hence sustainability can be built into the 'price'. As discussed above, SMA's design team – in consultation with the sustainability advisers – explored initiatives to meet the sustainability objectives, and devised practical targets for the reference design stage based on the information available to them. According to interviewee SD2, weekly sustainability design meetings or workshops were held.

Integration into planning approvals Environmental impact assessment reports were produced for both stage 1 and stage 2 of the project, under the Environmental Planning and Assessment Act 1979 (NSW). The environmental assessment reports summarised the sustainability initiatives. Some of these initiatives became part of the draft statement of commitments for stage 1, namely under design and construction stage commitments (Transport NSW, 2009a). Interviewee SS2 stated that the statement of commitments basically 'mapped what the designers were already doing'.

Any contractor selected to undertake planning, design, construction or operation of the CBD Metro would have been required to undertake all works in accordance with the final statement of commitments and the conditions of approval (Transport NSW, 2009a). Interviewee SS2 explained that 'what [SMA] did in the environmental assessment was have sustainability [objectives and initiatives] in the statement of commitments, because they get enshrined in law through the planning approval process'. Thus sustainability was effectively 'locked' into the project.

Integration into procurement The Sustainability Initiatives Register was provided to the organisations shortlisted to bid for the two key contracts. The Integrated Metro Operations (IMO) was to be a public-private partnership (PPP) and was for the rolling stock and rail systems, the track-work and power supply, station fit-out, and long term operations and maintenance. The Permanent Route Infrastructure (PRI), a design and construct contract, was for developing the tunnels, and station excavation and passenger access (Transport NSW, 2010b).

Detailed targets that met or exceeded those in the Sustainability Initiatives Register were expected to be developed for the construction stage and service stage of the IMO works; including for electricity consumption, fuel consumption, embodied energy of concrete and steel, and quantity of waste to be recycled (Transport NSW, 2010b).

Both contractors were expected to develop and submit a Sustainability

Plan that included processes and methodologies for embedding sustainability initiatives into their activities, as well as resources and staffing for sustainability (Transport NSW, 2009b). Interviewee SS2 believed that it was important that such a Sustainability Plan actually added value to the process, in particular in addition to the more traditional environmental management plans, to avoid sustainability practice being perceived as onerous or 'just best practice environmental management'.

Do – Organisational Structure, and Roles and Responsibilities SMA had a number of sustainability advisers working on the Sydney Metro project. There were sustainability advisers embedded in the design team, who worked in developing the Sustainability Initiatives Register and undertaking feasibility and other studies on potential sustainability initiatives. There was also an Environmental Sustainability Manager who was instrumental in developing the SMS. The sustainability advisers and Environmental Sustainability Manager reported to the Senior Manager Environment, and ultimately, the General Manager Station Precincts and Planning Approvals, and the Project Director. Interviewee SL1 observed that sustainability was treated as a separate but integrated discipline or work stream.

SMA had leadership-level support and commitment for sustainability which, as discussed by Lam et al. (2011) and Scanlon and Davis (2011), is paramount. SL2 believed that 'the leadership team is really critical. . . if you don't have anybody who supports [sustainability], then it will never get up'. Of equal importance was the advocacy and support role played by the sustainability advisers and manager. They worked across the different work streams, and especially in design and procurement, to ensure sustainability commitments were being taken seriously. The sustainability advisers were careful to use the same 'speak' or language as the discipline they were working with. Interviewee SS3 remarked that 'if you are talking about sustainability [the design team] can glaze over, but if you are talking about efficiency, then they totally understand what that means'. Further, interviewee SS2 discussed that it is important that the sustainability advisers had the appropriate technical and financial understanding in order to put cases together for the adoption of non-conventional initiatives. They then needed the tenacity to follow through embedding those initiatives into the various work streams. Underlying all of this was the need for enthusiasm and passion for sustainability.

Although there was no formal sustainability training programme, the sustainability advisers' advocacy role helped to build the awareness and capacity of the project team to address sustainability in their day-to-day activities – awareness that they could then take to other projects in legacy.

Check and Act – Measuring, Monitoring and Reporting Because the Sydney Metro project was deferred at a relatively early stage of delivery (at the letting of the IMO and PRI contracts), some of the monitoring and reporting processes were not wholly established. However, there were some plans in place.

SMA planned to measure, monitor and report progress against targets, and the targets against the policy positions and best practice. According to interviewee SS1, this process was to incorporate legislative and other changes in the revision of targets, and drive continual improvement in sustainability. Interviewee SL2 remarked that 'in the end what I expected to have was a sustainability assessment at the end of the design process which said ... "this is what you said you were going to do, this is what you did, [and] these are things you should try and do to improve the process next time"'.

Sustainability indicators through which to measure progress toward sustainability targets were to be negotiated during the process of procuring contractors. However it was noted by interviewee SS1 that the indicators would have been broadly consistent with the Australian Green Infrastructure Council's sustainability-rating tool once the tool was developed. Reporting on sustainability progress was to be on an annual basis consistent with the Global Reporting Initiative (2011) reporting protocol.

12.3.2 Olympic Delivery Authority's SMS

Plan – Sustainability Policy Commitments ODAT's sustainability policy commitments have stemmed from Olympic Delivery Authority and LOCOG commitments including to deliver 'the most sustainable Olympic Park ever' (Olympic Delivery Authority, 2007, p. 5).

The overarching sustainability governance documents for ODAT are the London 2012 Sustainability Policy and the London 2012 Sustainability Plan published in 2007. The London 2012 Sustainability Plan has a programme-wide commitment to sustainability across all venues and Games-related activities, and sets out activities through which to realise outcomes. The Olympic Delivery Authority then has a Sustainable Development Strategy that describes how it intends to deliver sustainability outcomes while complementing its other objectives including delivery on time and achieving value for money (Olympic Delivery Authority, 2007).

Plan – Sustainability Management System ODAT has developed an SMS to deliver on the commitments and requirements of the Sustainable Development Strategy and the Sustainable Transport Strategy (see below). The SMS is based on British Standard (BS) 8901:2009 Specification for a Sustainability Management System for Events,[7] which is currently being developed into an international standard ISO 20121. ODAT was one of the first organisations to achieve the revised standard (Lloyd's Register Quality Assurance, 2010).

ODAT's SMS is intentionally embedded in its operations, and forms part of an integrated Sustainability, Safety, Health, Environment and Quality Management System. It is therefore integrated with other ODAT processes including risk assessment and management, and stakeholder engagement (Olympic Delivery Authority Transport, 2010).

Plan – Prioritisation of Themes The Olympic Delivery Authority's Sustainable Development Strategy sets out ODAT's commitments across the five London 2012 headline sustainability themes including climate change, waste, biodiversity and ecology, inclusion, and healthy living. However, within ODAT's SMS the principal sustainability initiatives are focused under three of these five themes:

Healthy living Planning for 100 per cent of spectators to travel to all venues by public transport, walking and cycling. ODAT is working with Transport for London (TfL) to improve walking and cycling routes.

Inclusion A programme of accessibility for passengers with restricted mobility, and improving transport links from lower socio-economic areas of

[7]BS 8901 was introduced in Chapter 11.

London to the Games sites. It includes the provision of lifts in stations such as West Ham station, and audible and visual information facilities.

Climate change The Olympic Delivery Authority undertook a base-case carbon footprint study that included transport projects. The Orient Way project achieved a 23 per cent reduction in carbon against its 'base case' and an 'excellent' Civil Engineering Environmental Quality and Award Scheme (CEEQUAL) award – see CEEQUAL (2010a).

ODAT focused on these three themes as they were deemed most relevant to the transport programme of works. Focusing on three themes has also helped ODAT achieve 'bang for buck' and valuable outcomes within existing resources. The importance of prioritisation and materiality is discussed further in Chapter 11. Sustainability issues are also identified on a project case-by-case basis, as relevant to each of the five headline sustainability themes.

Do: Integration into Planning Approvals As outlined above, the principal statutory requirements for ODAT have come from the Olympic and Paralympic Games Act 2006. This act requires sustainable development to be addressed as part of the delivery of the Games, and that a Transport Plan be developed and kept under review. Environmental issues for ODAT were not determined by an environmental impact assessment, but rather by the strategic environment assessment (SEA) of the Transport Plan. The Environmental Plans and Programmes Regulations 2004 required that an SEA of ODAT's Transport Plan be undertaken. The SEA, and stakeholder consultation on the SEA Report, provided environmental issues input to the SMS.

However, rather than ODAT's sustainability agenda being based on outputs of the SEA, ODAT retained the aforementioned headline sustainability themes driven by the London 2012 Sustainability Policy and Sustainable Development Strategy as its formal sustainability objectives. The environmental issues arising from the SEA and other stakeholder interests have been managed and mitigated through the SMS – meaning that the SMS is not only used to manage non-statutory environmental and social issues encompassed by the sustainability objectives, but also to manage regulatory compliance issues arising through the planning process. The SMS is therefore a fundamental component of ODAT's overall management approach.

ODAT has also had to comply with a number of rail-related standards in the delivery of its rail infrastructure projects, including for West Ham, the London Underground standards for material compliance and the standard for planting of vegetation on railway embankments. ODAT has therefore had to ensure that any sustainability initiatives comply with these standards, an issue discussed further in section 12.4.1. Further, local authorities such as the London Borough of Newham have imposed planning conditions for each of ODAT's project sites. However, there has not been a direct link from the environmental impact assessment process for each project site and ODAT's sustainability objectives and targets. Again, sustainability activity on each project site has primarily been driven by the London 2012 sustainability commitments and managed through the SMS.

Do – Integration into other Project Delivery Activities The SMS has also allowed sustainability to be embedded in transport programme-related design, procurement and construction activities. This includes:

- Utilising contractual mechanisms and requiring contractors to use the CEEQUAL assessment tool and achieve a minimum rating of 'very good' – see CEEQUAL (2010b) – and to make every effort to achieve 'excellent';
- Considering contractor past sustainability performance in the tender evaluation process;
- Challenging suppliers to identify opportunities and to deliver sustainability enhancements and achieve greenhouse gas reductions through the design and construction phases;
- Taking a life-cycle approach through promotion of resource efficient material use and high re-use and recycling on- (or where not practicable, off-) site;
- Working closely with contractors, maintaining communication including on-site; and
- Operating a behavioural reward scheme with contractors (Whittaker et al., 2010).

Do – Organisational Structure, and Roles and Responsibilities The SMS clearly assigns roles and responsibilities for sustainability within ODAT, including at the leadership-level. For example, ODAT's Sustainability Manager is required to inform the Head of Transport Systems and Games Safety on sustainability risks and opportunities, who in turn authorises key decisions. ODAT's SMS is linked to a Learning and Development Programme for improved competence and training on sustainability. As part of the programme, existing and new employees are informed of their roles and responsibilities in relation to the SMS (Olympic Delivery Authority Transport, 2010).

ODAT's Sustainability Manager facilitates the integration of the sustainability objectives, targets and initiatives – and the SMS more broadly – with other systems, plans and procedures. For example, the Sustainability Manager draws from, and is required to comply with, the Communications Strategy and Code of Consultation produced by the Olympic Delivery Authority Communications Team when facilitating communications with internal and external stakeholders (Olympic Delivery Authority Transport, 2010).

ODAT's SMS has the commitment of the Executive Head of Directly Delivered Services who reports into the London 2012 Heads of Function. This has been important because according to interviewee OL2, 'your executives are generally the ones that will drive sustainability through the organisation. It is about having strong leadership at the top and true commitment'. Such leadership within ODAT also stems directly from the CEO of the Olympic Delivery Authority, who was viewed by interviewees OS2, OSL and OS3 as being passionate about achieving sustainability outcomes.

Check and Act – Reporting Procedures, Auditing and Management Reviews Each of ODAT's projects is required to report progress against sustainability objectives using project-specific sustainability performance indicators. This reporting feeds into the Olympic Delivery Authority's annual London 2012 Sustainability Plan Progress Report Card used to evaluate current and target sustainability performance. This provides a comprehensive account of progress on the commitments detailed in the London 2012 Sustainability Plan, and ODAT's performance on its sustainability objectives and targets is also tracked through this process.

The report cards are reviewed by the Commission for a Sustainable London 2012. The Commission monitors the sustainability performance of the whole of LOCOG, providing the external public assurance for the overall sustainability programme for the Games (Whittaker et al., 2010). The Commission produces periodic, independent reports on sustainability, which are made publicly available (Commission for a Sustainable London 2012, 2009).

The Commission also reviews aspects of ODAT's sustainability performance including through the report All Change – A Snapshot Review of Sustainability and Transport across the London 2012 Programme (Commission for a Sustainable London 2012, 2010). ODAT produces written Quarterly Transport Sustainability Reports and publishes numerous sustainability-related documents on the London 2012 website (LOCOG, 2011b).

In addition to adopting these reporting procedures, ODAT ensures compliance against its SMS and its sustainability commitments through second party assurance and compliance reviews through programmed sustainability audits. The agencies primarily responsible for the verification and audit are ODAT's internal audit teams and the Commission for a Sustainable London 2012.

Any issues are identified, reviewed and evaluated. Documented action plans are then put in place which detail the actions needed to amend the issues, and are then formally closed out with the responsible auditors and stakeholders (Olympic Delivery Authority Transport, 2010). The Head of Transport Systems and Games Safety is responsible for ensuring that management reviews of the SMS – including by ODAT's leadership-level – take place. This extensive system means that sustainability practice is constantly monitored and reported, with assurance being awarded where appropriate (Whittaker et al., 2010).

12.4 The Role of SMS in Overcoming Inhibitors

As highlighted above, some inhibitors to project sustainability were identified and discussed by the case study interviewees, and have also been identified in the literature and by the authors in their professional capacities. This section indicates how having an SMS can help overcome these inhibitors.

12.4.1 Overcoming Risk Aversion and Conservatism

A key inhibitor to sustainability is a culture of risk aversion and conservatism that exists within the built environment industry, particularly in disciplines such as procurement and design (Spangenberg et al., 2010). While there is increasing sustainability education in, for example, design disciplines (Schafer and Richards, 2007; Desha et al., 2007), Spangenberg et al. (2010, p. 1490) believe that sustainability is 'tangential to rather than embedded in mainstream design education and practice'.

Risk aversion is quite prevalent within the rail sector specifically. This is influenced by rail standardisation. As interviewee OD2 observed, 'there are a lot of standards in the railway that mandate you to do things a certain way and it can restrict what you do'. Interviewee OD2 added that these prescriptive engineering standards, such as for lighting, fire protection and safety, often do not allow the use of innovative technologies or materials.

An SMS, but more specifically the sustainability advisers and resources provided within that SMS, can help overcome such issues. Through researching the feasibility of initiatives, including through benchmarking (Scanlon and Davis, 2011), they can help re-define what is possible. Further, sustainability advisers' role in working across the different disciplines will help ensure consistent communication, advocacy and take up of the sustainability requirements and commitments.

12.4.2 Overcoming Perceived Technical, Cost and Time Constraints

Tan et al. (2011, p. 228) state that implementing sustainability practices can result in increases in cost and time, and that this 'discourages contractors from engaging actively in improving their sustainability performance'. However, often there is a perception as opposed to a reality of such constraints. This perception can also lead to the 'designing out' of sustainability initiatives in the name of budget savings (Pulaski et al., 2006), or in the name of simplifying the construction process (Pulaski and Horman, 2005).

Researching into possible sustainability initiatives through benchmarking and feasibility studies, as discussed above, can test assumptions and pre-conceptions about the costs of implementing sustainability initiatives. For example, the use of whole-of-life costing can help demonstrate a genuine business case for addressing sustainability (for good discussion on this issue, see Pearce (2008)). Further, having a mechanism built into an SMS to track progress, as well as locking sustainability targets and initiatives into conditions of approval, can help ensure that initiatives are not 'designed out'.

12.4.3 Balancing Decision Making Criteria

Although an SMS may enhance the level of knowledge about possible sustainability initiatives, a balance will still need to be achieved in decision making between sustainability, cost and time. For example in the context of design, interviewee OD1 observed that 'in the process of design there is no right or wrong answer, it's the best compromise that you come to between achievability and what you can get away with – with the neighbours, against what your greenhouse gas footprint might be, against what the cost might be... So it's this whole sort of compromise and there is a low point somewhere amongst all the requirements, which is where you hope to sit... '

An SMS can incorporate explicit but flexible rules about what constitutes acceptable decision making criteria – from the big decisions such as where to put a station or route alignment, to the smaller decisions about whether to use timber or steel on a handrail. An SMS can also help to give impetus to sustainability in decision making criteria because it should be accompanied by a clear policy commitment from the leadership-level, and because targets can be assigned to specific individuals and work streams with accountability for achieving them.

12.4.4 Understanding the 'True' Sustainability of Initiatives

Another inhibitor of sustainability in project delivery is a lack of understanding about the 'true' sustainability of a particular initiative. Demaid and Quintas (2006, p. 608) explain that 'industry has to accommodate to intended, unintended, rational and irrational effects when addressing sustainability issues', adding that 'the knowledge base is unstable'. They give a specific example of glass recycling and its energy intensity negating many perceived environmental benefits.

Similarly, Pearce et al. (2011, p. 1) talk about 'unintended negative system interactions'. They give an example of increased bacterial growth and reduction in water quality in water supply lines due to reduced flow rate from conserving water fixtures. In such cases, an initiative may achieve improved environmental performance but is ultimately unsustainable because it compromises the health, safety or quality of life of occupants, or the personnel who operate and maintain them (Pearce et al., 2011).

The aforementioned research into the viability of reaching certain targets and implementing certain initiatives can help address these issues. However, both the proponent and its delivery partners need to have access to information on the real financial and overall performance of sustainability initiatives, including of alternative materials. This will be influenced by the availability and quality of information in the public domain. In terms of information on materials, this is improving in Australia through various projects of the Building Products Innovation Council and other such organisations. Where there is good information, and through life-cycle and other assessments of key initiatives and material options, potentially negative consequences and trade-offs (Kemp et al., 2005) can be revealed.

12.4.5 Integrating Impact Assessment Outputs, Sustainability and Project Delivery

An SMS can make the important links between the impact assessment outputs, sustainability objectives and targets, and other project delivery activities. For example, in the case of SMA, provisions in the planning approval process were used to give mandatory status to sustainability targets and initiatives relating to detailed design, procurement and construction that had already been developed by the project team as part of planning (including concept design) processes. This highlights the potential benefits of project planning being sufficiently well-progressed by the time the impact assessment is undertaken and the approvals documentation submitted, to enable concrete and substantial sustainability commitments to be made at this point.

A lack of, or poor, enforcement of conditions of approval can lessen the effectiveness of using the planning approval process in this manner. Enforcement will depend on how objective or clear the conditions of approval are – if they are too broad or subjective, measuring, monitoring and enforcement may prove difficult or near impossible. Enforcement will also depend on the legislative robustness of the enforcement powers and the capacity, including resources, of the enforcement agency. The extent to which specific legislation permits consideration of broader sustainability issues within impact assessment processes is also a key consideration here.

Further, the impact assessment process can help those responsible for sustainability and the SMS to identify early in the life of the project which sustainability issues and opportunities may be present, which are most important, and hence which should be prioritised. This is still the case for ODAT where, though its sustainability objectives are a result of Games-related sustainability policy commitments, the SEA process revealed environmental issues that have also been managed through the SMS. An SMS can also assist in the planning approval process, especially by communicating to key stakeholders, including regulators and the community, that sustainability is being taken seriously, which may streamline the planning approval process. An SMS should therefore be explicitly referred to in impact assessment documentation.

Importantly, the SMS can take the impact assessment outputs and approval commitments and make explicit links to other delivery activities. This can include by allocating responsibility for addressing certain sustainability issues and objectives, such as reducing resource use, to the appropriate discipline or work-stream, such as to structural engineers and procurement personnel. An SMS can therefore address the issue of impact assessment practitioners often having limited access to decision makers in these other disciplines or work-streams. Additional sustainability issues and opportunities are likely to arise after the approvals process as the project progresses. Therefore, the SMS should be flexible enough to enable emerging issues to be effectively incorporated into project delivery activities, as well as facilitate innovation and creativity in responding to them.

12.5 Conclusion

This chapter has highlighted two case study SMSs – one from SMA and the other from ODAT. This chapter has also discussed how these SMSs can help overcome inhibitors to embedding sustainability in various project delivery phases and activities. The documentation of these approaches is important for sharing experiences with the broader industry both within and outside of Australia and the UK. This kind of knowledge sharing is critical for driving a step change within industry.

SMA's SMS comprised a number of key components including its Sustainability Policy; a team of sustainability advisers; and locking in sustainability objectives, targets and initiatives into design, planning approval and procurement documentation. This level of sophistication would very likely have resulted in sustainability outcomes during the delivery of the Sydney Metro project, albeit that this cannot be confirmed as the project was deferred by the State Government in 2010.

ODAT's SMS is based on BS 8901. Its primary purpose is to address the sustainability policy commitments laid out by the Olympic Delivery Authority and other Games partners including LOCOG; however ODAT has focused on three key themes of healthy living, inclusion and climate change that are deemed most relevant to the transport programme. The SMS is integrated into ODAT's Sustainability, Safety, Health, Environment and Quality Management System, and explicit links have been made to design, procurement and construction activities, including through the specification of the achievement of a CEEQUAL rating. A post-Games review of the success of the SMS in achieving sustainability outcomes in project delivery will be of value, and this may take place through the Olympic Delivery Authority's planned Legacy Learning Programme.

In both case studies, the SMSs provided a structured framework to facilitate a consistent and meaningful commitment to sustainability throughout the life of the project or project programme. This helped to ensure that potential negative impacts and risks were minimised, while opportunities were realised, to ultimately deliver real sustainability outcomes.

Epilogue

The intent of the book is to collect, present, and critically analyse the effective management of environmental (and other types of) impacts throughout the life of projects – as much in an analytic or theoretical way, as through practice. This effort comes to strengthen the relatively slow evolution of the core idea. The way forward may still have to be discovered, as there are many options worth exploring.

As in most innovative work, paving the way forward needs hard work and special qualities. Before anything else, it needs good preparation, which points to studying across the fields – in this case, at least EIA and EMS. It also needs good and brave ideas, which points towards creativity and the confidence to support any 'unusual' hypotheses. More experimentation is also necessary, similar to the ways described here, or perhaps exploring complementary patterns and issues. And, finally, paving the way forward also needs effective communication and dissemination, which points to networking and publications.

The panorama of the impact continuum appears to have an interesting combination of traits regarding human resources. For a start, there is the classic duel between academics and practitioners – as should be expected. Then there is a special separation: that of the EIA camp versus the EMS camp, which appears to traverse training and practice: each project involves further but related groups of staff within different stages of the project life cycle – for example, those involved in project planning and those who operate the project once constructed. And then there is a changing scene of 'who-is-who': impact continuum specialists seem to be dropping out after a few years of work, which is at least curious – if not indicating a kind of a 'dead-end' or a long battle beyond one's productive lifetime. At the same time, others join in, which seems to indicate that there is still some interest.

All kinds of professionals are essential for the necessary innovation in the impact continuum, in different and complementary functions: academics for preparing, reporting, and criticising it, practitioners for applying it, and proponents for supporting it. At a larger scale, there is also a wider group of 'environmental' professionals who also need to make their precious contribution: those involved in the environmental standards – ranging from international authorities to interest groups. Environmental standards must be coordinated, at least for being coherent, easy to access, working together for a common objective, and showing the way to forthcoming standards.

A. Perdicoúlis, B. Durning, and L. Palframan
December 2011

Bibliography

Ahmadvand M., E. Karami, G.H. Zamani, and F. Vanclay (2009), Evaluating the use of Social Impact Assessment in the context of agricultural development projects in Iran, *Environmental Impact Assessment Review*, **29**(6): 399–407.

American Society for Quality (2011), *Plan, Do, Check, Act Cycle*, `http://asq.org/learn-about-quality/project-planning-tools/overview/pdca-cycle.html` Accessed 03 August 2011.

Argote L., B. McEvily, and R. Reagans (2003), Managing knowledge in organizations: An integrative framework and review of emerging themes, *Management Science*, **49**:571–582.

Arts, J. (1998), *EIA Follow-up – on the role of ex-post evaluation in Environmental Impact Assessment*, Groningen: GeoPress.

Arts, J., P. Caldwell, and A. Morrison-Saunders (2001), Environmental Impact Assessment Follow-up: Good Practice and Future Directions – Findings from a Workshop at the IAIA 2000 Conference, *Impact Assessment and Project Appraisal*, **19**(3):175–185.

Arts, J., and A. Morrison-Saunders (2004), Theoretical perspectives on EIA and follow-up. In: A. Morrison-Saunders and J. Arts (eds) *Impact Assessment: Handbook of EIA and SEA Follow-up*, London: Earthscan.

Arts, J., and S. Nooteboom (1999), Environmental impact assessment monitoring and auditing. In: J. Petts (ed.) *Handbook of Environmental Impact Assessment (Vol. 1)*, Oxford: Blackwell Science.

Australian Government Department of Agriculture, Fisheries and Forestry (2009), *Australia's National Framework for Environmental Management Systems in Agriculture*, `http://www.daff.gov.au/natural-resources/land-salinity/ems/framework` Accessed 02 November 2011.

Australian Green Infrastructure Council (2009), *Fact Sheet No 2 – AGIC Infrastructure Sustainability Assessment Categories*, `www.agic.net.au/`

fact_sheet_2_agic_infra_sustainability_assessment_categories.pdf Accessed 03 August 2011.

Australian Green Infrastructure Council (2011), *Pilot Rating Tool*, www.agic.net.au Accessed 03 August 2011.

Ayande Independent Iranians Online Press Media (2011), *30 fatal accidents in 17 month: breaking record in Iran oil and gas industry* (in Persian), http://www.ayandenews.com/news/34382/ Dated: 10 August 2011; Accessed: 29 August 2011.

Azapagic, A. (2003), Systems approach to corporate sustainability: a general management framework, *Process Safety and Environmental Protection*, **81**(5):303–316.

Bailey, J. (1997), Environmental Impact Assessment and Management: An Underexplored Relationship, *Environmental Management*, **21**(3):317–327.

Baker, J. (2004), A practical framework for EIA and follow-up. In: A. Morrison-Saunders and J. Arts (eds) *Impact Assessment: Handbook of EIA and SEA Follow-up*, London: Earthscan.

Barnes, J., and D. Lemon (1999), Life-of-project environmental management strategy: Case study of the confederation Bridge project, Canada, *Journal of Environmental Assessment Policy and Management*, **1**(4):429–439.

Beanlands, G., and P. Duinker (1984), An Ecological Framework for Environmental Impact Assessment, *Journal of Environmental Management*, **18**: 267–277.

Benn, S., D. Dunphy, and A. Griffiths (2006), Enabling Change for Corporate Sustainability: An Integrated Perspective, *Australasian Journal of Environmental Management*, **13**:156–165.

Bond, A., C.V. Viegas, C.C.S.R. Coelho, and P.M. Selig (2010), Informal knowledge processes: the underpinning for sustainability outcomes in EIA? *Journal of Cleaner Production*, **18**:6–13.

Boswell, P. (2005), Project Sustainability Management: A Systems Approach, www.docstoc.com/docs/64071832/Systems-Approach-to-Project-Management Accessed 10 January 2010.

British Petroleum (2010), *Deep Water Horizon Accident Investigation Report*, http://www.bp.com/liveassets/bp_internet/globalbp/globalbp_uk_english/gom_response/STAGING/local_ assets/downloads_pdfs/Deepwater_Horizon_Accident_Investigation_Report.pdf Accessed 21 November 2011.

British Standards Institute (2010), BS 8901 Sustainability Management Systems for Events, www.bsigroup.co.uk/en/Assessment-and-Certification-services/Management-systems/Standards-and-Schemes/BS-8901/ Accessed 09 May 2011.

Broderick, M. (2011), Leading by example – The Equator Principles are allowing low-income economies to lead the way in ESIA practice, *Mining, People and the Environment*, July 2011.

Broderick, M., and B. Durning (2006), Environmental impact assessment and environmental management plans – an example of an integrated process from the UK. In: J.F. Martin-Duque, C.A. Brebbia, D. Emmanouloudis and U. Mander (eds) *Geo-Environment and Landscape Evolution II WIT Transactions Ecology and the Environment*, Volume 89, Southampton: WITPress.

Broderick, M., and B. Durning (2010), Follow-up in ESIA as an aid to Greening Economies, *Proceedings of the 30th Conference of the International Association for Impact Assessment*, Geneva.

Broderick, M., B. Durning, and D. Ferguson (2010), Equator principles and the minerals industry. *Proceedings of the 30th Annual Conference of the International Association for Impact Assessment*, Geneva: Switzerland.

Brouwer, M., and C.S.A. van Koppen (2008), The soul of the machine: continual improvement in ISO 14001, *Journal of Cleaner Production*, **16**(4):450–457.

Buntaine, M.T. (2011), Does the Asian Development Bank respond to past environmental performance when allocating environmentally risky financing? *World Development*, **39**(3):336–350.

Cagno, E., A. DiGiulio, and P. Trucco (1999), A methodological framework for the initial environmental review (IER) in EMS implementation. *Journal of Environmental Assessment Policy and Management*, **1**(4):505–532.

Canter, L.W. (1998), Methods for Effective Environmental Information Assessment (EIA) Practice. In: A.L. Porter and J. Fittipaldi (eds) *Environmental Methods Review: Retooling Impact Assessment for the New Century*, Atlanta: AEPI.

Carroll, B., and T. Turpin (2002), *Environmental Impact Assessment Handbook: a Practical Guide for Planners, Developers and Communities*, London: Thomas Telford.

Carruthers, G., and G. Tinning (2003), Where, and How, Do Monitoring and Sustainability Indicators Fit into Environmental Management Systems, *Australian Journal of Experimental Agriculture*, **43**:307–323.

Caspary, G. (2009), Assessing, mitigating and monitoring environmental risks of large infrastructure projects in foreign financing decisions: the case of OECD-country public financing for large dams in developing countries, *Impact Assessment and Project Appraisal*, **27**(1):19–32.

CEEQUAL (2010a), *Assessment Manual for Projects in the UK and Ireland* (Version 4.1), www.ceequal.com/downloads.htm Accessed 03 March 2011.

CEEQUAL (2010b), *About Awards*, www.ceequal.co.uk/awards_how.htm Accessed 02 February 2010.

CEN – European Committee for Standardization (2004), *CWA 14924-5 – European Guide to Good Practice in Knowledge Management – Part 5: KM Terminology*, http://www.cen.eu/cen/Sectors/Sectors/ISSS/CEN%20Workshop%20Agreements/Pages/Knowledge%20Management.aspx Accessed 06 January 2012

Cherp, A. (2008), The role of environmental management systems in enforcing standards and thresholds in the context of EIA follow-up. In: M. Schmidt et al. (eds) *Standards and Thresholds for Impact Assessment*, Berlin: Springer-Verlag.

Communities and Local Government (2004), *Planning policy statement 23: planning and pollution control*, London: HMSO.

Commission for a Sustainable London 2012 (2009), *Procuring a Legacy*, www.cslondon.org/wp-content/uploads/downloads/2009/01/2009_ODA_Procurement_Review.pdf Accessed 26 June 2010.

Commission for a Sustainable London 2012 (2010), *All Change*, www.cslondon.org/wp-content/uploads/downloads/2010/06/CSL_Transport_Review.pdf Accessed 25 May 2011.

Dao, M.A., and G. Ofori (2010), Determinants of firm compliance to environmental laws: a case study of Vietnam, *Asia Europe Journal*, **8**(1):91–112.

Dao, M.A., G. Ofori, and L.S. Pheng (2009), Benefits of Compliance With Local Environmental Regulations: MNCs' Perspectives. In: D. Meijer and F.D. Jong (eds) *Environmental Regulation: Evaluation, Compliance and Economic Impact*, Hauppauge NY: Nova Science Publishers.

Dao, M.A., G. Ofori, and S.P. Low (2011), *Determinants of Firm Compliance with Environmental Laws: A Case Study of Vietnam*. Singapore: National University of Singapore, PhD Thesis.

Defra (2007a) *Waste strategy for England 2007*, London: DEFRA.

Defra (2007b), *Incineration of municipal solid waste*, London: DEFRA.

Defra (2007c), *Planning and pollution control: Improving the way the regimes work together in delivering new development*, London: DEFRA.

Defra (2010), *Statistical release: survey of industrial and commercial waste arisings 2010: final results*, http://www.defra.gov.uk/evidence/statistics/environment/waste/documents/stats-release101216.pdf Accessed 16 February 2011.

Defra (2011a), *Local authority collected waste generation, UK*, http://www.defra.gov.uk/statistics/environment/waste/wrfg19-munwaste/ Accessed 17 June 2011.

Defra (2011b), *Municipal waste management in the European Union*, http://www.defra.gov.uk/statistics/environment/waste/wrfg08-munec/ Accessed 16 June 2011

Demaid, A., and P. Quintas (2006), Knowledge Across Cultures in the Construction Industry: Sustainability, Innovation and Design, *Technovation*, **26**(5–6):603–610.

Department for Transport Highways Agency (2011), *Design Manual for Roads and Bridges (DMRB)*, http://www.dft.gov.uk/ha/standards/dmrb/index.htm Accessed 30 October 2011.

Desha, C.J.K., K. Hargroves, M.H. Smith, and P. Stasinopoulos (2007), The Importance of Sustainability in Engineering Education: A Toolkit of Information and Teaching Material, *Proceedings of the Engineering Training and Learning Conference*, www.naturaledgeproject.net/Documents/ICDPaper-Final.pdf Accessed 03 June 2011.

DETR (2000), *Waste Strategy 2000, England and Wales (Part 1)*, London: HMSO.

Devuyst, D. (1999), Sustainability Assessment: The Application of a Methodological Framework, *Journal of Environmental Assessment, Policy and Management*, **1**(4):459–487.

Dias, E.G.C.S., and L.E. Sánchez, (2000) Environmental impact assessment: evaluating the follow- up phase. In: R.K. Singhal and A.K. Mehrotra (eds) *Environmental Issues and Management of Waste in Energy and Mineral Production*, Rotterdam: Balkema.

Doppelt, B. (2003), *Leading Change Towards Sustainability: A Change Management Guide for Business, Government and Civil Society*, Sheffield: Greenleaf Publishing.

Duffy, P., and R. DuBois (1999), *Environmental Impact Guidelines – What are Environmental Assessments?*, `http://www.fao.org/docrep/007/x4005e/x4005e00.htm` Accessed 21 November 2011.

Dunphy, D., A. Griffiths, and S. Benn (2003), *Organisational Change for Corporate Sustainability*, London: Routledge.

EBRD (2008), *Environmental and Social Policy*, London: European Bank for Reconstruction and Development.

Eccleston, C. (1998), A Strategy for Integrating NEPA with EMS and ISO 14000, *Environmental Quality Management*, **7**(3):9–17.

Eccleston, C.H., and R.B. Smythe (2002), Integrating Environmental Impact Assessment with Environmental Management Systems, *Environmental Quality Management*, **11**(4):1–13.

Emilsson, S., and O. Hjelm (2009), Towards sustainability management systems in three Swedish local authorities, *Local Environment*, **14**(8):721–732.

ENS (2001), Marshlands of The Tigris-Euphrates Delta 90 Percent Gone, *Environmental News Services*, `http://ens-news.com/ens/may2001/2001L-05-18-01.html` Accessed 04 May 2010.

Environment Agency (2011a), *How to comply with your Environmental Permit (EPR 1.00)*, Bristol: Environment Agency.

Environment Agency (2011b), *Application for an environmental permit (Part B2 – General – new bespoke permit)*, Bristol: Environment Agency.

Environment Agency (2011c), *Opra for EPR version 3.6. Annex B – Opra scheme for waste facilities*, Bristol: Environment Agency.

Environmental Resources Management (2004), *The interaction between Land Use Planning and Environmental Regulation*, Research Findings No.192/2004. Edinburgh: Scottish Executive.

E&P-Forum (1994), *Guidelines for the Development and Application of Health, Safety and Environmental Management Systems*, E&P forum. Report No. 6.36/210.

E&P-Forum/UNEP (1997), *Environmental Management in Oil and Gas Exploration and Production: An overview of issues and management approaches*, Joint E&P (Exploration and Production) Forum/UNEP Technical Publication. E&P Forum Report 2.72/254; UNEP IE/PAC Technical Report 37.

EPD – Environmental Protection Department (2002), *The Operation of the EIA Ordinance*, Hong Kong: EPD.

Equator Principles (2011), *Equator Principles*, `http://www.equator-principles.com/documents/Equator_Principles.pdf` Accessed October 2011.

Esquer-Peralta, J., L. Velaquez,, and N. Munguia (2009), Perceptions of core elements for sustainability management systems, *Management Decision*, **46**(7):1027–1038.

European Commission (2001), Regulation (EC) No 761/2001 of the European Parliament and of the Council, of 19 March 2001, allowing voluntary participation by organisations in a Community eco-management and audit scheme (EMAS), *Official Journal of the European Communities*, **L114**, 24.4.2001.

European Commission (2008), *Directive 2008/98/EC of the European Parliament and of the Council of 19 November 2008 on waste and repealing certain Directives*, Brussels: European Commission.

European Communities (2011), *The European Eco-Management and Audit Scheme. Improving your environmental and business performance*, Brussels: European Communities.

Faith-Ell, C., and J. Arts (2009), Public Private Partnerships and EIA: Why PPP are Relevant to Practice of Impact Assessment for Infrastructure, *Proceedings of the 29th Annual Conference of the International Association for Impact Assessment*, Accra.

Ferguson, D. (2008), *The Equator Principles: Improving EIA Practice? An Assessment of the Impact of the Equator Principles on Public Engagement and Impact Mitigation and Monitoring in Developing Economies*, Oxford: Oxford Brookes University, MSc Dissertation.

Fernández-Sánchez, G., and F. Rodríguez-López (2010), A Methodology to Identify Sustainability Indicators in Construction Project Management – Application to Infrastructure Projects in Spain, *Ecological Indicators*, **10**(6):1193–1201.

General Statistics Office of Vietnam (2006), *Statistical Handbook of Vietnam, 2006*, http://www.gso.gov.vn/default_en.aspx?tabid=515&idmid=5&ItemID=5910

Genskow, K.D., and D.M. Wood (2011), Improving voluntary environmental management programs: facilitating learning and adaptation, *Environmental Management*, **47**:907–916.

GHK (2008), *Evaluation on EU legislation – Directive 85/337/EEC (Environmental impact assessment, EIA) and associated amendments; Final report submitted by GHK, Technopolis within the framework of ENTR/04/093-FC-Lot 1*, http://ec.europa.eu/environment/eia/pdf/Evaluation%20of%20EIA.pdf

Gibson, R., S. Hassan, S. Holtz, J. Tansey, and G. Whitelaw (2005), *Sustainability Assessment: Criteria and Processes*, London: Earthscan.

Gilpin, A. (ed.) (1996), *Dictionary of Environment and Sustainable Development*, Chichester: John Wiley & Sons.

Glasson, J., R. Therivel, and A. Chadwick (2005), *Introduction to Environmental Impact Assessment* (3rd edn), London: Routledge.

Global Reporting Initiative (2011), *Sustainability Reporting Guidelines – Version 3.1*, www.globalreporting.org/NR/rdonlyres/53984807-9E9B-4B9F-B5E8-77667F35CC83/0/G31GuidelinesinclTechnicalProtocolFinal.pdf Accessed 04 June 2011.

Goodland, R., and J.R. Mercier (1999), *The Evolution of Environmental Assessment in the World Bank: from 'Approval' to Results*, The World Bank, Environment Department, Paper 67.

Grant Thornton (2008), *An ever-changing landscape. Waste sector M&A analysis, Summer 2008*, London: Grant Thornton.

Haas, P. (1992), Introduction: epistemic communities and international policy coordination, *International Organization*, **46**(1):1–35.

Hacking, T., and P. Guthrie (2008), A Framework for Clarifying the Meaning of Triple Bottom-Line, Integrated, and Sustainability Assessment, *Environmental Impact Assessment Review*, **28**(1):73–89.

Harden, M., and N. Walker (eds) (2004), *Lesson not learned – The other Shell reports*, http://www.foe.co.uk/resource/reports/lessons_not_learned.pdf Accessed 21 November 2011.

Harmer, C. (2005), *Is Improving the Effectiveness of Environmental Impact Assessment in the UK Dependent on the Use of Follow-up? Views of environmental consultants*, Norwich: University of East Anglia, Masters Thesis.

Harris, L.C., and A. Crane (2002), The Greening of Organisational Culture: Management Views on the Depth, Degree and Diffusion of Change, *Journal of Organisational Change Management*, **15**(3):214–234.

Harris, R., and A. Khare (2002), Sustainable development issues and strategies for Alberta's oil industry, *Technovation*, **22**(9):571–583.

Hickie, D., and M. Wade (1997), The Development of Environmental Action Plans: Turning Statements into Actions, *Journal of Environmental Planning and Management*, **40**(6):789–801.

Holling, C.S. (1978), *Adaptive Environmental Assessment and Management*, Chichester: Wiley & Sons.

Hong Kong Environmental Protection Department (1998), *Guidelines for Development Projects in Hong Kong – Environmental Monitoring and Audit*, Hong Kong: HKEPD.

IAIA (1999), *Principles of EIA Best Practice*, `http://www.iaia.org/publicdocuments/special-publications/` Accessed 14 November 2011.

IAIA (2006), Public participation: international best practice principles, *IAIA Special Publication Series*, No.4, Fargo: International Association for Impact Assessment.

IAIA (2009), *What is Impact Assessment?*, Fargo: International Association for Impact Assessment.

Iberdrola (2011), *Iberdrola surpasses 8,000 megawatts of renewable capacity in Europe*, `http://www.iberdrola.es/webibd/corporativa/iberdrola?IDPAG=ENMODPRENNAC11&URLPAG=/gc/prod/en/comunicacion/notasprensa/110810_NP_01_MegavatiosRenovables.html` Accessed 21 September 2011.

ICMM (2009), *Good Practice Guidance for Mining and Biodiversity*, London: The International Council on Mining and Metals.

ICOFC-West (2009), *Introduction to Iranian Central Oil Field Company, West Operational Area* (Persian Text), `http://www.gharb.icofc.org/index.aspx?siteid=78&pageid=379` Accessed 02 August 2011.

IEMA (2004), Guidelines for Environmental Impact Assessment, *IEMA Per-*

spectives, Lincoln: IEMA.

IEMA (2008), Environmental management plans, *Best practice series*, Vol. 12, IEMA, Lincoln: IEMA.

IEMA (2010), *IEMA Acorn Scheme*, http://www.iema.net/ems/acorn_scheme Accessed: 02 November 2011.

IEMA (2011), *Special Report – State of Environmental Impact Assessment in the UK*, www.iema.net Accessed October 2011.

IFC (2006), *Performance Standards on Social and Environmental Sustainability*, http://www.ifc.org/ifcext/sustainability.nsf/Content/PerformanceStandards Accessed October 2011.

IFC (2011), *Environmental, Health and Safety Guidelines*, http://www.ifc.org/ifcext/sustainability.nsf/Content/EHSGuidelines Accessed: October 2011.

ILO-OSH (2001), *Guidelines on occupational safety and health management systems*, Geneva: International Labor Office.

Independent Public Inquiry (2010), *Long-Term Public Transport Plan for Sydney*, www.transportpublicinquiry.com.au/pdf/F2e_Public_Transport_Inquiry_Final_Report_26May2010_Chapter_1_Community_views.pdf Accessed 30 June 2010.

IPIECA/API/OGP (2005), *Oil and Gas Industry Guidance on Voluntary Sustainability Reporting: Using Environmental, Health & Safety, Social and Economic Performance Indicators* (1st ed.), The International Petroleum Industry Environmental Conservation Association (IPIECA), American Petroleum Institute (API), International Association of Oil & Gas Producers (OGP).

IPIECA/API/OGP (2010), *Oil and gas industry guidance on voluntary sustainability reporting, The global oil and gas industry association for environmental and social issues* (2nd edn), The American Petroleum Institute (API), International Association of Oil & Gas Producers (OGP).

ISO 14031 (1999), *Environmental Management – Guideline for Environmental Performance Evaluation (EPE) ISO 140031:1999*, Geneva: International Organization for Standardization.

ISO (2004), *International Standard ISO 14001 (2nd edition) Environmental Management Systems – Requirements with Guidance for Use*, Geneva: International Organization for Standardization.

ISO (2009), *International Standard ISO 14050: Environmental Management – Vocabulary* (3rd edn), Geneva: International Organization for Standardization.

ISO (2010a), *ISO 9001 certifications top one million mark, food safety and information security continue meteoric increase*, `http://www.iso.org/iso/pressrelease.htm?refid=Ref1363` Accessed 02 November 2011.

ISO (2010b), *ISO to develop sustainable event standard in run-up to 2012 Olympics*, `http://www.iso.org/iso/pressrelease.htm?refid=Ref1281` Accessed 24 October 2011.

Japanese Ministry of the Environment (2002), *Overseas Environmental Measures of Japanese Companies (Vietnam)*, `http://www.env.go.jp/earth/coop/oemjc/viet/e/contents.html` Accessed October 2011.

Jha-Thakur, U., T.B. Fischer, and A. Rajvanshi (2009), Reviewing design stage of environmental impact assessment follow-up: looking at the open cast coal mines in India, *Impact Assessment and Project Appraisal*, **27**(1):33–44.

Johnson, D., and C. Walck, (2004), Integrating Sustainability into Corporate Management Systems, *Journal of Forestry*, **102**(5):32–39.

Jørgensen, T.H., A. Remmen, and M.D. Mellado (2006), Integrated Management Systems: Three Different Levels of Integration, *Journal of Cleaner Production*, **14**(8):713–722.

Kagioglou, M., R. Cooper, G. Aouad, and M. Sexton (2000), Rethinking Construction: The Generic Design and Construction Process Protocol, *Engineering, Construction and Architectural Management*, **7**(2):141–153.

Kemp, R., S. Parto, and R.B. Gibson (2005), Governance for Sustainable Development: Moving from Theory to Practice, *International Journal of Sustainable Development*, **8**(1/2):12–30.

Khabaronline (2011), *17 months incident record of Iranian oil, gas and petrochemical industry: 30 people lost their lives and more than 100 people badly inquired* (Persian Text), `http://www.khabaronline.ir/news-167214.aspx` Accessed 29 August 2011.

Lam, P.T.I., E.H.W. Chan, C.K. Chau, S.C. Poon, and K.P. Chun (2011), Environmental Management System vs Green Specifications: How Do they Complement Each Other in the Construction Industry? *Journal of Environ-*

mental Management, **92**:788–795.

Lave, J., and E. Wenger (1991), *Situated Learning – Legitimate Peripheral Participation*, Cambridge: Cambridge University Press.

Le, H.N. (2010), Development of EIA system in Vietnam, 1993–2010. *Proceedings of the 3rd Meeting of Signatories to SEA Protocol and 14th Meeting of Working Group on EIA*, Geneva, 22–26 November.

Le, T.C. (1997), Development of Environmental Impact Assessment in Vietnam, *Proceedings of the First Workshop in Training in Environmental Impact Assessment*, Hanoi: Vietnam.

Le, Q.T., and A.N. Nguyen (2001), *Wastewater Management in Industrial Estates in Vietnam*, University for Agricultural and Forestry Study, Ho Chi Minh City, Vietnam.

Le, D.A., T.C. Le, H. Luc, and N.N. Sinh (2000a), *General Guideline Book for EIA of Development Projects, Building for Environmental Management in Vietnam*, Hanoi: Ministry of Natural Resources and Environment.

Le, D.A., T.C. Le, H. Luc, and N.N. Sinh (2000b), *General Guideline Book for EIA of Tourism Development Projects, Building for Environmental Management in Vietnam*, Hanoi: Ministry of Natural Resources and Environment.

Le, D.A., T.C. Le, H. Luc, L.H. Ke, and N.N. Sinh (2000c), *General Guidebook for EIA of Urban Planning, Building for Environmental Management in Vietnam*, Hanoi: Ministry of Natural Resources and Environment.

Liew, A. (2007), Understanding data, information, knowledge, and their interrelationships, *Journal of Knowledge Management Practice*, **8**(2) – no pagination.

Lloyd's Register Quality Assurance (2010), *ODA Breaks New Ground with Sustainability Award for 2012 Transport Plans*, `www.lrqa.co.uk/news/Client/oda.aspx` Accessed 05 January 2010.

LOCOG (2011a), *The London Organising Committee of the Olympic Games and Paralympic Games Limited (LOGOC)*, `www.london2012.com/about-us/the-people-delivering-the-games/the-olympic-delivery-authority/` Accessed 15 June 2011.

LOCOG (2011b), *Official Website of the London 2012 Summer Olympic Games and Paralympic Games*, `www.london2012.com/` Accessed 21 July 2011.

Lohani, B.N., W.J. Evans, R.R. Everitt, H. Ludwig, R.A. Carpenter, and S.L. Tu

(1997), *Environmental Impact Assessment for Developing Countries in Asia*, Manila: Asian Development Bank.

Luc, H., and D.A. Le (2000), Summary and conclusion, *Proceedings of the Second Workshop on Environmental Impact Assessment, Building for Environmental Management in Vietnam*, Hanoi, Vietnam, 23 September 1998.

Lundberg, K. (2011), A systems thinking approach to environmental follow-up in a Swedish Central Public Authority: hindrances and possibilities for learning from experience, *Environmental Management*, **48**:123–133.

MacDonald, J.P. (2005), Strategic sustainable development using the ISO 14001 standard, *Journal of Cleaner Production*, **13**(6):631–643.

Marshall, R. (2002), Professional practice: Developing environmental management systems to deliver mitigation and protect the EIA process during follow-up, *Impact Assessment and Project Appraisal*, **20**(4):286–292.

Marshall, R. (2004), Can industry benefit from participation in EIA-follow-up? – the ScottishPower experience. In: A. Morrison-Saunders and J. Arts (eds) *Assessing Impact: Handbook of EIA and SEA Follow-up*, London: Earthscan.

Marshall, R. (2005), Environmental impact assessment follow-up and its benefits for industry, *Impact Assessment and Project Appraisal*, **23**(3):191–196.

Marshall, R., and A. Morrison-Saunders (2003), EIA Follow-up – linking impact assessment with implementation, *The Environmentalist*, **17**:16–19.

Marshall, R., A. Morrison-Saunders, and J. Arts (2005), International principles for best practice EIA follow-up, *Impact Assessment and Project Appraisal*, **23**(3):175–181.

May R.F. (2003), Marine conservation reserves, petroleum exploration and development, and oil spills in coastal waters of Western Australia, *Marine Pollution Bulletin*, **25**(5–8):147–154.

McKillop J., and A.L. Brown (1999), Linking project appraisal and development: the performance of EIA in large scale mining projects, *Journal of Environmental Assessment Policy and Management*, **1**(4):99–428.

Melnyk, S.A., R.P. Sroufe, and R. Calantone (2003), Assessing the impact of environmental management systems or corporate and environmental performance, *Journal of Operations Management*, **21**:329–351.

Mordue, M. (2008), *The Monitoring of Mitigation Measures in Environmental Impact Assessment*, Oxford: Oxford Brookes University, MSc Dissertation.

Morris, P., and, R. Therivel (2009), *Methods of Environmental Impact Assess-*

ment (3rd ed.), Oxford: Taylor & Francis.

Morrison-Saunders, A., J. Baker, and J. Arts (2003), Lesson from practice: towards successful follow-up, *Impact Assessment and Project Appraisal*, **21**(1):43–56.

Morrison-Saunders, A., and J. Arts (2004a), Introduction to EIA follow-up. In: A. Morrison-Saunders and J. Arts (eds) *Assessing Impact: Handbook of EIA and SEA Follow-up*, London: Earthscan.

Morrison-Saunders, A., and J. Arts (2004b), Exploring the Dimensions of EIA Follow-up, *Proceedings of 24th annual meeting of the International Association for Impact Assessment*, Vancouver, Canada.

Morrison-Saunders, A., and J. Arts (2005), Learning from experience: emerging trends in environmental impact assessment follow-up – Editorial, *Impact Assessment and Project Appraisal*, **23**(3):170–174.

Morrison-Saunders, A., R. Marshall, and J. Arts. (2007), EIA Follow-Up International Best Practice Principles, *Special Publication Series No.6*, Fargo: International Association for Impact Assessment.

Moser T. (2001), MNCs and Sustainable Business Practice: The Case of the Colombian and Peruvian Petroleum Industries, *World Development*, **29**(2):291–309.

Nawrocka, D., and T. Parker (2009), Finding the connection: environmental management systems and environmental performance, *Journal of Cleaner Production*, **17**(6):601–607.

NEPA (1969), *US National Environmental Policy Act*, `http://epw.senate.gov/nepa69.pdf` Accessed October 2011.

Netherwood, A. (1994), Environmental management systems. In: R. Welford (ed.) *Corporate Environmental Management: Systems and Strategies*, London: Earthscan.

Nitz, T., and L. Brown (2002), Piecing together the jigsaw of environmental management tools, *Proceedings of the 22nd Annual Meeting of the International Association for Impact Assessment*, Hague.

Noble, B. (2002), The Canadian experience with SEA and sustainability, *Environmental Impact Assessment Review*, **22**(1):3–16.

Noble, B.F. (2010), *Introduction to Environmental Impact Assessment: Guide to Principles and Practice* (2nd ed.), Toronto: Oxford University Press.

Noble, B., and K. Storey (2005), Towards increasing the utility of follow-up in

Canadian EIA, *Environmental Impact Assessment Review*, **25**(2):163–180.

Office of the Commissioner for Environmental Sustainability – Organisational Structures and Sustainability (2011), www.ces.vic.gov.au/CA256F310024B628/0/F9DC51ECC67A00D9CA25785E00132986/$File/Organisational+Structures+and+Sustainability.pdf Accessed 03 August 2011.

OGP (2010), *Environmental performance in the E&P industry – 2009 data*, Report No.442, London: International Association of Oil & Gas Producers.

OHSAS (2007), *18001: Occupational Health and Safety Management Systems – Requirements*, Fife: OHSAS.

Olympic Delivery Authority (2007), *Sustainable Development Strategy*, www.london2012.com/publications/sustainable-development-strategy-full-version.php Accessed 17 April 2011.

Olympic Delivery Authority (2010), *London 2012 Transport – On Track*, www.london2012.com/publications/london-2012-transport-on-track.php Accessed 17 April 2011.

Olympic Delivery Authority Transport (2010), *SSHEQ Management System – Procedure S2.25: Sustainability Management System* (not publicly available)

Oskarsson, K., and F. von Malmborg (2005), Integrated Management Systems as a Corporate Response to Sustainable Development, *Corporate Social Responsibility and Environmental Management*, **12**:121–128.

OSPAR Commission (2003), *OSPAR Recommendation 2003/5 to Promote the Use and Implementation of Environmental Management Systems by the Offshore Industry*, London: OSPAR Commission.

Palframan, L., J.O. Jenkins, and X. Zhang (2010), *A Role for Corporate Sustainability Strategy in the Garden City*, www.inter-disciplinary.net/wp-content/uploads/2010/06/Palframan-paper.pdf Accessed 23 May 2011.

Palframan, L. (2010), The integration of environmental impact assessment and environmental management systems: experiences from the UK, *Proceedings of the 30th Annual Meeting of the International Association for Impact Assessment*, Geneva.

Pearce, A.R., A.P. McCoy, T. Smith-Jackson, M. Edwards, A. Pruden, and R. Bagchi (2011), Designing a Sustainable Built Environment: A Hazards Analysis and Critical Control Points (HACCP) Approach, *Proceedings of Engineering Sustainability 2011*, Pittsburgh.

Pearce, A.R. (2008), Sustainable Capital Projects: Leapfrogging the First Cost Barrier, *Civil Engineering and Environmental Systems*, **25**(4):291–300.

Perdicoúlis, A. (2010), *Systems Thinking and Decision Making in Urban and Environmental Planning*. Cheltenham, UK and Northampton, MA, USA: Edward Elgar.

Perdicoúlis, A. (2011), *Building Competences for Spatial Planners: Methods and Techniques for Performing Tasks with Efficiency*. London: Routledge.

Perdicoúlis, A., and B. Durning (2007), An alternating-sequence conceptual framework for EIA–EMS integration, *Journal of Environmental Assessment Policy and Management*, **9**(4):385–397.

Perdicoúlis, A., and B. Durning (2009), An alternating-sequence conceptual framework for EIA–EMS integration. In: W.R. Sheate (ed.) *Tools, Techniques and Approaches for Sustainability*, Singapore: World Scientific.

Petts, J. (1999), *Handbook of Environmental Impact Assessment: Environmental Impact Assessment: Process, Methods and Potential*, London: Wiley–Blackwell.

Petts, J. (1994), Incineration as a waste management option. In: R.E. Hester and RM Harrison (eds) *Waste Incineration and the Environment*, Cambridge: Royal Society of Chemistry.

Pope, J. (2006), Editorial: What's So Special About Sustainability Assessment?, *Journal of Environmental Assessment, Policy and Management*, **8**(3):v–ix.

Pope, J., and W. Grace (2006), Sustainability Assessment in Context: Issues of Process, Policy and Governance, *Journal of Environmental Assessment, Policy and Management*, **8**(3):373–398.

Provincial Government of the Western Cape (2005), *Guidelines for Environmental Management Plans*, `http://www.asapa.org.za/images/uploads/8_deadp_emp_guideline_june05.pdf` Accessed November 2011.

Pulaski, M.H., and M.J. Horman (2005), Continuous Value Enhancement Process, *Journal of Construction Engineering and Management*, **131**(12):1274–1282.

Pulaski, M.H., M.J. Horman, and D.R. Riley (2006), Constructability Practices to Manage Sustainable Building Knowledge, *Journal of Architectural Engineering*, **12**(2):83–92.

Rankin, M., B. Griffin, and M. Broderick (2008), IFC raises the bar on environmental health and safety, *International Mining*, **April**.

Rees C. (1999), Improving the effectiveness of environmental assessment at the World Bank, *Environmental Impact Assessment Review*, **19**(3):333–339.

Ridgway, B. (1999), The project cycle and the role of EIA and EMS, *Journal of Environmental Assessment Policy and Management*, **1**(4):393–405.

Ridgway, B. (2005), Environmental management system provides tools for delivering on environmental impact assessment commitments, *Impact Assessment and Project Appraisal*, **23**(4):325–331.

Runhaar, H. (2011), 25 years of Environmental Impact Assessment in the Netherlands Effectively governing environmental protection? *Proceedings of 'Health in Environmental Assessment: Developing an International Perspective'*, University of Liverpool, June 28th.

Saarikoski, H. (2000), Environmental impact assessment (EIA) as collaborative learning process, *Environmental Impact Assessment Review*, **20**(6):681–700.

Sadler, B. (1996), *Environmental assessment in a changing world: evaluating practice to improve performance. A report for the International Study of the Effectiveness of Environmental Assessment*, Canadian Environmental Assessment Agency and International Association for Impact Assessment.

Salomone, R. (2008), Integrated Management Systems: Experiences in Italian Organizations, *Journal of Cleaner Production*, **16**(16):1786–1806.

Sánchez, L.E. (2006), Environmental impact assessment and its role in project management. In: A. Vilela Jr. and J. Demajorovic (eds) *Models and Tools for Environmental Management* (in Portuguese), São Paulo: Ed. Senac.

Sánchez, L.E., and A.L.C.F. Gallardo (2005), On the successful implementation of mitigation measures, *Impact Assessment and Project Appraisal*, **23**(3):182–190.

Sánchez, L.E., and T. Hacking (2002), An approach to linking environmental impact assessment and environmental management systems, *Impact Assessment and Project Appraisal*, **20**(1):25–38.

Sánchez, L.E., and A. Morrison-Saunders (2011), Learning about knowledge management for improving environmental impact assessment in a government agency: the Western Australian experience, *Journal of Environmental Management*, **92**:2260–2271.

Sanvicens, G., and P. Baldwin (1996), Environmental monitoring and audit in Hong Kong, *Journal of Environmental Planning & Management*, **39**(3):429–440.

Sarkis, J., and R. Sroufe (eds) (2007), *Strategic Sustainability: The State of the Art in Corporate Environmental Management Systems*, Sheffield: Greenleaf Publishing.

Scanlon, J., and A. Davis (2011), The Role of Sustainability Advisers in Developing Sustainability Outcomes for an Infrastructure Project: Lessons from the Australian Urban Rail Sector, *Impact Assessment and Project Appraisal*, **29**(2):121–133.

Scanlon J., and S. Hodgson (2010), Delivering urban rail sustainably: Rail Express – Critical Issues for Rail's Key Areas, *Informa Australia*, **2**:48–50.

Schafer, A., and B.S. Richards (2007), From Concept to Commercialisation: Student Learning in a Sustainable Engineering Innovation Project, *European Journal of Engineering Education*, **32**(2):143–165.

Scott, W.R. (1995), *Institutions and Organizations*, Thousand Oaks, CA: Sage.

Scottish Government (2011), *Planning Circular 3, 2011: The Town and Country Planning (Environmental Impact Assessment) (Scotland) Regulations 2011*, Edinburgh: Scottish Government.

ScottishPower Renewables (2011), *ScottishPower Renewables Becomes the First Developer in the UK to Achieve 1 GW of Capacity From Wind Power*, `http://www.scottishpowerrenewables.com/pages/press_releases.asp?article=108&date_year=2011` Accessed 21 September 2011.

Scrase, J.I., and W.R. Sheate (2002), Integration and integrated approaches to assessment: What do they mean for the environment? *Journal of Environmental Policy & Planning*, **4**:275–294.

Sebastiania, M., E. Martin, D. Adrianza, C. Méndez, M. Villaróa, and Y. Saud (2001). Linking impact assessment to an environmental management system. Case study: a downstream upgrading petroleum plant in Venezuela, *Environmental Impact Assessment Review*, **21**(2):137–168.

Sheate, W.R. (1999), Editorial, *Journal of Environmental Assessment Policy and Management*, **1**(4):iii–v.

Sheate, W.R. (2002), Workshop on 'Linking impact assessment and management tools' – Conference report, *Journal of Environmental Assessment Policy and Management*, **4**(4):465–474.

Shelbourn, M.A., D.M. Bouchlaghem, C.J. Anumba, P.M. Carillo, M.M.K. Kahlfan, and J. Glass (2006), Managing Knowledge in the Context of Sustainable Construction, *Journal of Information Technology in Construction*, **11**:57–71.

Shell (2004), *Impact Assessment Yellow Book*, Shell Internationale Petroleum Maatschappij B.V./Royal Dutch Shell Group.

Shen, L.Y., J.L. Hao, V.W.Y. Tam, and H.Yao (2007), A Checklist for Assessing Sustainability Performance of Construction Projects, *Journal of Civil Engineering and Management*, **13**(4):273–281.

Sinclair, A.J., A. Diduck, and P. Fitzpatrick (2008), Conceptualizing learning for sustainability through environmental assessment: critical reflections on 15 years of research, *Environmental Impact Assessment Review*, **28**(7):415–428.

Slinn, P., J. Handley, and S.A. Jay (2007), Connecting EIA to environmental management systems: lessons from industrial estate developments in England, *Corporate Social Responsibility and Environmental Management*, **14**(2):88–102.

Spangenberg, J.H., A. Fuad-Luke, and K. Blincoe (2010), Design for Sustainability (DfS): The Interface of Sustainable Production and Consumption, *Journal of Cleaner Production*, **18**:1485–1493.

Suff, P. (2011), Quality & environment: 2 standards – 1 system, *IEMA Magazine*, **March**:30–32.

Tan, Y., L.Shen, and H.Yao (2011), Sustainable Construction Practice and Contractors' Competitiveness: A Preliminary Study, *Habitat International*, **35**:225–230.

Tenev, S., A. Carlier, O. Chaudry, and Q.T. Nguyen (2003), *Informality and the Playing Field in Vietnam's Business Sector*, Washington, D.C.: International Finance Corporation.

The Highland Council (2010), *Construction Environmental Management Process for Large Scale Projects*, `http://www.highland.gov.uk/NR/rdonlyres/485C70FB-98A7-4F77-8D6B-ED5ACC7409C0/0/construction_environmental_management_22122010.pdf` Accessed November 2011.

Thérivel, R. (2004), *Strategic Environmental Assessment in Action*, London: Earthscan.

Thérivel, R., and M.R. Partidário (1996), *The Practice of Strategic Environmental Assessment*, London: Eathscan.

Tinker, L., D. Cobb, A. Bond, and M. Cashmore (2005), Impact mitigation in environmental impact assessment: paper promises or the basis of consent decisions?, *Impact Assessment and Project Appraisal*, **23**(4):265–280.

Tinsley, S., and I. Pillai (2006), *Environmental Management Systems: Understanding Organizations Drivers and Barriers*, London: Earthscan.

Tran, T.T.P. (1996), Environmental Management and Policy Making in Vietnam. Australian National University, `http://coombs.anu.edu.au/~vern/env_dev/papers/pap03.html` Accessed August 2003.

Tran, V. Y. et al. (2000), The basis for the establishment of a general EIA guideline book, appropriate to Vietnamese conditions, *Proceedings of the Third Workshop on Environmental Impact Assessment, building for Environmental Management in Vietnam*, Hanoi, Vietnam.

Transport NSW (Sydney Metro Authority) (2009a), *Sydney Metro Network Stage 1 (CBD Metro) Environmental Assessment*, `www.transport.nsw.gov.au/sites/default/file/metrodocs/` Accessed 16 April 2011.

Transport NSW (Sydney Metro Authority) (2009b), *Design and construction of PRI – 1 works PRI-1 contract exhibit: A scope of works and technical criteria appendix 24 – project plan requirements*, `www.transport.nsw.gov.au/sites/default/file/metrodocs/` Accessed 23 September 2010.

Transport NSW (Sydney Metro Authority) (2010a), *Sydney Metro Network Stage 2 (Central – Westmead) draft environmental assessment report*, `www.transport.nsw.gov.au/sites/default/file/metrodocs/` Accessed 23 May 2011.

Transport NSW (Sydney Metro Authority) (2010b), *Sydney Metro – background information*, `www.transport.nsw.gov.au/sites/default/file/metrodocs/` Accessed 16 April 2011.

Turpin, T. (2010), International study of the Effectiveness of Environmental Assessment Update 2010, *EIA Effectiveness IEMA Workshop*, March 2010, Swindon.

Ugochukwu, C.N.C., and J. Ertel (2008), Negative impacts of oil exploration on biodiversity management in the Niger Delta area of Nigeria, *Impact Assessment and Project Appraisal*, **26**(2):139–147.

Ugwu, O.O., M.M. Kumaraswamy, A. Wong, and S.T. Ng (2006), Sustainability Appraisal in Infrastructure Projects (SUSAIP): Part 1. Development of Indicators and Computational Methods, *Automation in Construction*, **15**(2):239–251.

UNDP (1995), *Incorporating Environmental Considerations into Investment decision making in Vietnam*, Hanoi: UNDP.

UNECA (2009), *Africa Review Report on Waste Management*, Addis Ababa: UNECA.

UNEP (2002), *Topic 11 – EIA Implementation and follow-up in the EIA Training*, Resource Manual, United Nations Environmental Programme (UNEP).

US EPA – United States Environmental Protection Agency (1989), *Fundamentals of Environmental Compliance Inspections*, Rockville: Government Institutes.

US EPA – United States Environmental Protection Agency (1992), *Principles of Environmental Enforcement*, Washington DC: Environmental Protection Agency.

Vanclay, F. (2004), The triple bottom line and impact assessment: How do TBL, EIA, SIA, SEA and EMS relate to each other? *Journal of Environmental Assessment Policy and Management*, **6**(3):265–288.

Vanegas, J.A. (2003), Road Map and Principles for Built Environment Sustainability, *Environmental Science and Technology*, **37**(23):5363–5372.

Varnäs, A., C. Faith-Ell, and B.. Balfors (2009), Linking environmental impact assessment, environmental management systems and green procurement in construction projects: lessons from the City Tunnel Project in Malmö, Sweden, *Impact Assessment and Project Appraisal*, **27**(1):69–76.

Vietnam Fatherland Front (2008), `http://www.mattran.org.vn/` Accessed March 2008.

Viridor (2010), *Viridor corporate responsibility report 2010*, Taunton: Viridor.

Viridor (2011), *Viridor's Business Management Systems – providing an Integrated Management System: Informing customers, employees, the community and other stakeholders*, Taunton: Viridor.

Walmsley, B., and P. Tarr (2011), *Quality Assurance in EIA: Guide and Review,*

Conference Training Course, 30th Annual Conference of the International Association for Impact Assessment (IAIA), Puebla Mexico.

Wardell Armstrong (2010), *Viridor: Re-phasing, vertical and lateral extension with enhanced restoration and ecological scheme at Pilsworth South Landfill Site – Volume 2: Environmental Statement*, Stoke on Trent: Wardell Armstrong.

WBCSD – World Business Council for Sustainable Development (1996), *Environmental Assessment: A Business Perspective*, Geneva: WBCSD.

Werr, A., and T. Stjernberg (2003), Exploring management consulting firms as knowledge systems, *Organization Studies*, **24**(6):881–908.

Whittaker, S., S. Wilson, A. Davis, J. Scanlon, and A. Rode (2010), *The Legacy of the Green Games – Creating New Benchmarks in Infrastructure Sustainability*, `www.manidisroberts.com.au/events/2010/infrastructure-sustainability/thought-piece.html` Accessed 10 December 2010.

Wilson, M., and R. Pearce (2009), *The UK's top twenty*, Chartered Institution of Wastes Management, **October**:20–25.

Wood, C. (2003), *Environmental Impact Assessment: a Comparative Review* (2nd ed.), Harlow: Pearson Education.

Wood, C., and J. Bailey (1994), Predominance and independence in environmental impact assessment: The Western Australia model, *Environmental Impact Assessment Review*, **14**:37–59.

Wood, G. (2008), Thresholds and criteria for evaluating and communicating impact significance in environmental statements: 'see no evil, hear no evil, speak no evil'?, *Environmental Impact Assessment Review*, **28**(1):22–38.

Wood, G., A. Rodriguez-Bachiller, and J. Becker (2007), Fuzzy sets and simulated environmental change: evaluating and communicating impact significance in environmental impact assessment, *Environment and Planning A*, **39**:810–829.

World Bank (1993), *The World Bank and Environmental Assessment: An Overview*, `http://siteresources.worldbank.org/INTSAFEPOL/1142947-1116495579739/20507374/Update1TheWorldBankAndEAApril1993.pdf` Accessed November 2011.

World Bank, (1999a), *Pollution prevention and abatement handbook*

1998: toward cleaner production (Report No.19128), http://www-wds.worldbank.org Accessed October 2011.

World Bank (1999b), *Environmental Management Plans*, http://siteresources.worldbank.org/INTSAFEPOL/1142947-1116495579739/20507392/Update25EnvironmentalManagementPlansJanuary1999.pdf Accessed November 2011.

World Bank (2011), *Operating Manual: Operating Policy 4.01 – Environmental Assessment*, http://web.worldbank.org/ Accessed October 2011.

Index